Are Cell Phones Dangerous?
A Look at Mobile Phone Radiation and Your Health

Contents

Chapter 1

Introduction

1.1 Mobile phone

"Cell Phone" redirects here. For the film, see Cell Phone (film).
"Handphone" redirects here. For the film, see Handphone (film).

Evolution of mobile phones, through early smartphone

A **mobile phone** (also known as a **cellular phone**, **cell phone**, **hand phone**, or simply a **phone**) is a phone that can make and receive telephone calls over a radio link while moving around a wide geographic area. It does so by connecting to a cellular network provided by a mobile phone operator, allowing access to the public telephone network. By contrast, a cordless telephone is used only within the short range of a single, private base station.

In addition to telephony, modern mobile phones also support a wide variety of other services such as text messaging, MMS, email, Internet access, short-range wireless communications (infrared, Bluetooth), business applications, gaming, and photography. Mobile phones that offer these and more general computing capabilities are referred to as smartphones.

The first hand-held cell phone was demonstrated by John F. Mitchell[1][2] and Martin Cooper of Motorola in 1973, using a handset weighing around 4.4 pounds (2 kg).[3] In 1983, the DynaTAC 8000x was the first to be commercially available. From 1983 to 2014, worldwide mobile phone subscriptions grew from zero to over 7 billion, penetrating 100% of the global population and reaching the bottom of the economic pyramid.[4] In 2014, the top cell phone manufacturers were Samsung, Nokia, Apple, and LG.[5]

1.1.1 History

Main article: History of mobile phones
A hand-held mobile radiotelephone is an old dream of radio engineering. In 1917, Finnish inventor Eric Tigerstedt filed a patent for what he described as a "pocket-size folding telephone with a very thin carbon microphone". Among other early descriptions is one found in the 1948 science fiction novel *Space Cadet* by Robert Heinlein. The protagonist, who has just traveled to Colorado from his home in Iowa, receives a call from his father on a telephone in his pocket. Before leaving for earth orbit, he decides to ship the telephone home "since it was limited by its short range to the neighborhood of an earth-side [i.e. terrestrial] relay office." Ten years later, an essay by Arthur C. Clarke envisioned a "personal transceiver, so small and compact that every man carries one." Clarke wrote: "the time will come when we

1

Martin Cooper of Motorola made the first publicized handheld mobile phone call on a prototype DynaTAC model on April 4, 1973. This is a reenactment in 2007.

will be able to call a person anywhere on Earth merely by dialing a number." Such a device would also, in Clarke's vision, include means for global positioning so that "no one need ever again be lost." In his 1962 *Profiles of the Future*, he predicted the advent of such a device taking place in the mid-1980s.[6]

Early predecessors of cellular phones included analog radio communications from ships and trains. The race to create truly portable telephone devices began after World War II, with developments taking place in many countries. The advances in mobile telephony have been traced in successive *generations* from the early "0G" (zeroth generation) services like the Bell System's Mobile Telephone Service and its successor, Improved Mobile Telephone Service. These "0G" systems were not cellular, supported few simultaneous calls, and were very expensive.

The first handheld mobile cell phone was demonstrated by Motorola in 1973. The first commercial automated cellular network was launched in Japan by Nippon Telegraph and Telephone in 1979. In 1981, this was followed by the simultaneous launch of the Nordic Mobile Telephone (NMT) system in Denmark, Finland, Norway and Sweden.[7] Several other countries then followed in the early to mid-1980s. These first generation ("1G") systems could support far

The Motorola DynaTAC 8000X. First commercially available, hand-held cellular mobile phone, 1984

more simultaneous calls, but still used analog technology.

In 1991, the second generation (2G) *digital* cellular technology was launched in Finland by Radiolinja on the GSM standard. This sparked competition in the sector as the new operators challenged the incumbent 1G network operators.

Ten years later, in 2001, the third generation (3G) was launched in Japan by NTT DoCoMo on the WCDMA standard.[8] This was followed by 3.5G, 3G+ or turbo 3G enhancements based on the high-speed packet access (HSPA) family, allowing UMTS networks to have higher data transfer speeds and capacity.

By 2009, it had become clear that, at some point, 3G networks would be overwhelmed by the growth of bandwidth-intensive applications like streaming media.[9] Consequently, the industry began looking to data-optimized fourth generation technologies, with the promise of speed improvements up to 10-fold over existing 3G technologies. The first two commercially available technologies billed as 4G were the WiMAX standard (offered in North America by Sprint) and the LTE standard, first offered in Scandinavia by TeliaSonera.

1.1.2 Features

Main article: Mobile phone features
See also: Smartphone

All mobile phones have a number of features in common, but manufacturers also seek product differentiation by adding functions to make them more attractive to consumers. This has led to great innovation in mobile phone development over the past 20 years.

The common components found on all phones are:

- A battery, providing the power source for the phone functions.

- An input mechanism to allow the user to interact with the phone. The most common input mechanism is a keypad, but touch screens are also found in most smartphones.

- A screen which echoes the user's typing, displays text messages, contacts and more.

- Basic mobile phone services to allow users to make calls and send text messages.

- All GSM phones use a SIM card to allow an account to be swapped among devices. Some CDMA devices also have a similar card called a R-UIM.

- Individual GSM, WCDMA, iDEN and some satellite phone devices are uniquely identified by an International Mobile Equipment Identity (IMEI) number.

Low-end mobile phones are often referred to as feature phones, and offer basic telephony. Handsets with more advanced computing ability through the use of native software applications became known as smartphones.

Several phone series have been introduced to address specific market segments, such as the RIM BlackBerry focusing on enterprise/corporate customer email needs; the Sony-Ericsson 'Walkman' series of music/phones and 'Cyber-shot' series of camera/phones; the Nokia Nseries of multimedia phones, the Palm Pre the HTC Dream and the Apple iPhone.

Sound quality

In sound quality, smartphones and feature phones vary little. Some audio-quality enhancing features like Voice over LTE and HD Voice have appeared and are often available on newer smartphones. Sound quality can remain a problem with both, as this depends, not so much on the phone itself, as on the quality of the network, and in case of long distance calls, the bottlenecks/choke points met along the way.[10][11] As such, on long-distance calls even features such as Voice over LTE, HD voice may not improve things. In some cases smartphones can improve audio quality even on long-distance calls, by using VoIP phone service, with someone else's WiFi/internet connection.[12]

Text messaging

Main article: SMS

The most commonly used data application on mobile phones is SMS text messaging. The first SMS text message was sent from a computer to a mobile phone in 1992 in the UK, while the first person-to-person SMS from phone to phone was sent in Finland in 1993.

The first mobile news service, delivered via SMS, was launched in Finland in 2000, and subsequently many organizations provided "on-demand" and "instant" news services by SMS.

Multimedia Messaging Service (MMS) was introduced in 2001.

SIM card

Main articles: Subscriber Identity Module and Removable User Identity Module
GSM feature phones require a small microchip called a Subscriber Identity Module or SIM card, to function. The SIM card is approximately the size of a small postage stamp and is usually placed underneath the battery in the rear of

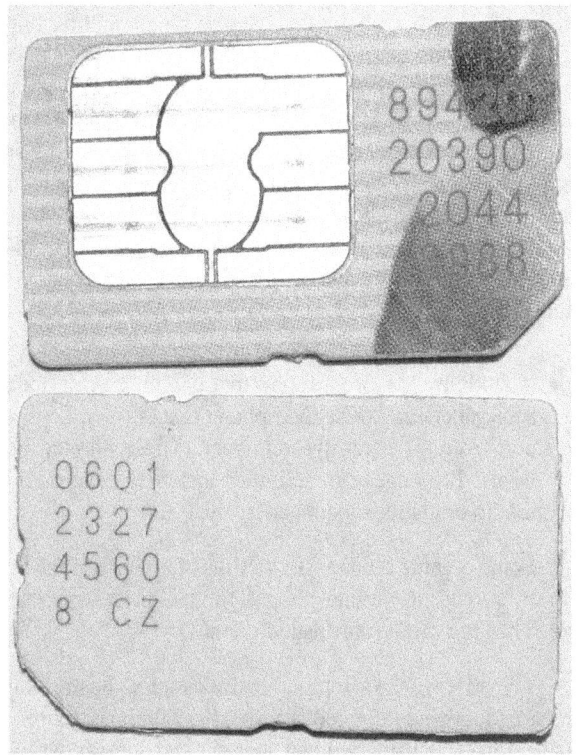

Typical mobile phone SIM card

Kosher phones

There are Jewish orthodox religious restrictions which, by some interpretations, standard mobile telephones do not meet. To solve this issue, some rabbinical organizations have recommended that phones with text messaging capability not be used by children.[15] These restricted phones are known as kosher phones and have rabbinical approval for use in Israel and elsewhere by observant Orthodox Jews. Although these phones are intended to prevent immodesty, some vendors report good sales to adults who prefer the simplicity of the devices. Some phones are even approved for use by essential workers (such as health, security and public services) on the sabbath, even though use of any electrical device is generally prohibited.[16]

1.1.3 Mobile phone operators

1980

Growth in mobile phone subscribers per country from 1980 to 2009.

Main article: Mobile phone operator

The world's largest individual mobile operator by subscribers is China Mobile with over 500 million mobile phone subscribers.[17] Over 50 mobile operators have over 10 million subscribers each, and over 150 mobile operators had at least one million subscribers by the end of 2009.[18] In 2014, there were more than seven billion mobile phone subscribers worldwide, a number that is expected to keep growing.

1.1.4 Manufacturers

See also: List of best-selling mobile phones and List of mobile phone makers by country

Prior to 2010, Nokia was the market leader. However, since then competition emerged in the Asia Pacific region with brands such as Micromax, Nexian, and i-Mobile and chipped away at Nokia's market share. Android powered smartphones also gained momentum across the region at the expense of Nokia. In India, their market share also dropped

the unit. The SIM securely stores the service-subscriber key (IMSI) and the K_i used to identify and authenticate the user of the mobile phone. The SIM card allows users to change phones by simply removing the SIM card from one mobile phone and inserting it into another mobile phone or broadband telephony device, provided that this is not prevented by a SIM lock.

The first SIM card was made in 1991 by Munich smart card maker Giesecke & Devrient for the Finnish wireless network operator Radiolinja.

Multi-card hybrid phones

A hybrid mobile phone can hold up to four SIM cards. SIM and RUIM cards may be mixed together to allow both GSM and CDMA networks to be accessed.

From 2010 onwards they became popular in India and Indonesia and other emerging markets,[13] attributed to the desire to obtain the lowest on-net calling rate. In Q3 2011, Nokia shipped 18 million of its low cost dual SIM phone range in an attempt to make up lost ground in the higher end smartphone market.[14]

significantly to around 31 percent from 56 percent in the same period. Their share was displaced by Chinese and Indian vendors of low-end mobile phones.[19]

In Q1 2012, based on Strategy Analytics, Samsung surpassed Nokia, selling 93.5 million units and 82.7 million units, respectively. Standard & Poor's has also downgraded Nokia to 'junk' status at BB+/B with negative outlook due to high loss and still declined with growth of Lumia smartphones was not sufficient to offset a rapid decline in revenue from Symbian-based smartphones over the next few quarters.[20]

In Q3 2014, the top 10 manufacturers were Samsung (20.6%), Nokia (9.5%), Apple Inc. (8.4%), LG (4.2%), Huawei (3.6%), TCL Communication (3.5), Xiaomi (3.5%), Lenovo (3.3%), ZTE (3.0%) and Micromax (2.2%).[21]

- Note: Vendor shipments are branded shipments and exclude OEM sales for all vendors

Other manufacturers outside the top five include TCL Communication, Lenovo, Sony Mobile Communications, Motorola. Smaller current and past players include Karbonn Mobile, Audiovox (now UTStarcom), BenQ-Siemens, BlackBerry, Casio, CECT, Coolpad, Fujitsu, HTC, Just5, Kyocera, Lumigon, Micromax Mobile, Mitsubishi Electric, Modu, NEC, Neonode, Openmoko, Panasonic, Palm, Pantech Wireless Inc., Philips, Qualcomm Inc., Sagem, Sanyo, Sharp, Sierra Wireless, SK Teletech, Soutec, Trium, Toshiba, and Vidalco.

1.1.5 Use of mobile phones

General

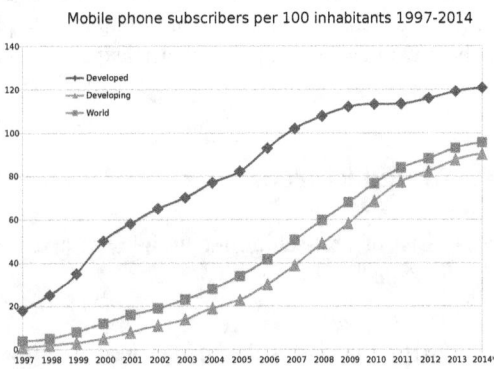

Mobile phone subscribers per 100 inhabitants 1997-2014

Mobile phone subscribers per 100 inhabitants. 2014 figure is estimated.

Mobile phones are used for a variety of purposes, including keeping in touch with family members, conducting business, and having access to a telephone in the event of an emergency. Some people carry more than one cell phone for different purposes, such as for business and personal use. Multiple SIM cards may also be used to take advantage of the benefits of different calling plans—a particular plan might provide cheaper local calls, long-distance calls, international calls, or roaming. The mobile phone has also been used in a variety of diverse contexts in society, for example:

- A study by Motorola found that one in ten cell phone subscribers have a second phone that is often kept secret from other family members. These phones may be used to engage in activities including extramarital affairs or clandestine business dealings.[24]

- Some organizations assist victims of domestic violence by providing mobile phones for use in emergencies. They are often refurbished phones.[25]

- The advent of widespread text messaging has resulted in the cell phone novel; the first literary genre to emerge from the cellular age via text messaging to a website that collects the novels as a whole.[26]

- Mobile telephony also facilitates activism and public journalism being explored by Reuters and Yahoo![27] and small independent news companies such as Jasmine News in Sri Lanka.

- The United Nations reported that mobile phones have spread faster than any other technology and can improve the livelihood of the poorest people in developing countries by providing access to information in places where landlines or the Internet are not available, especially in the least developed countries. Use of mobile phones also spawns a wealth of micro-enterprises, by providing work, such as selling airtime on the streets and repairing or refurbishing handsets.[28]

- In Mali and other African countries, people used to travel from village to village to let friends and relatives know about weddings, births and other events, which is now avoided within mobile phone coverage areas, which are usually more extensive than land line penetration.

- The TV industry has recently started using mobile phones to drive live TV viewing through mobile apps, advertising, social tv, and mobile TV.[29] 86% of Americans use their mobile phone while watching TV.

- In parts of the world, mobile phone sharing is common. It is prevalent in urban India, as families and groups of friends often share one or more mobiles among their members. There are obvious economic benefits, but often familial customs and traditional gender roles play a part.[30] It is common for a village to have access to only one mobile phone, perhaps owned by a teacher or missionary, but available to all members of the village for necessary calls.[31]

Smartphones

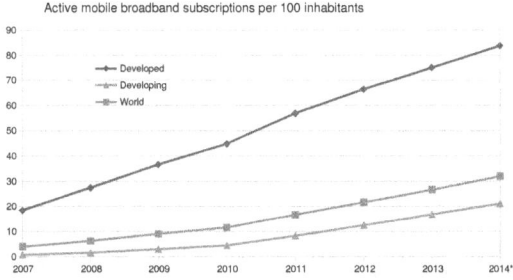

Active mobile broadband subscriptions per 100 inhabitants.[32]

Smartphones have a number of distinguishing features, but the ITU measures those with internet connection which it calls *Active Mobile-Broadband subscriptions* (which includes tablets etc.). In the developed world, these have now overtaken the usage of earlier mobile systems. However, in the developing world, they account for only 20%.

For distributing content

In 1998, one of the first examples of distributing and selling media content through the mobile phone was the sale of ringtones by Radiolinja in Finland. Soon afterwards, other media content appeared such as news, video games, jokes, horoscopes, TV content and advertising. Most early content for mobile tended to be copies of legacy media, such as the banner advertisement or the TV news highlight video clip. Recently, unique content for mobile has been emerging, from the ringing tones and ringback tones in music to "mobisodes", video content that has been produced exclusively for mobile phones.

In 2006, the total value of mobile-phone-paid media content exceeded Internet-paid media content and was worth 31 billion dollars.[33] The value of music on phones was worth 9.3 billion dollars in 2007 and gaming was worth over 5 billion dollars in 2007.[34]

While driving

Main article: Mobile phones and driving safety
Mobile phone use while driving is common but controver-

A sign along Bellaire Boulevard in Southside Place, Texas (Greater Houston) states that using mobile phones while driving is prohibited from 7:30 AM to 9:30 AM and from 2:00 PM to 4:15 PM

sial. Being distracted while operating a motor vehicle has been shown to increase the risk of accidents. Because of this, many jurisdictions prohibit the use of mobile phones while driving. Egypt, Israel, Japan, Portugal and Singapore ban both handheld and hands-free use of a mobile phone; others —including the UK, France, and many U.S. states— ban handheld phone use only, allowing hands-free use.

Due to the increasing complexity of mobile phones, they are often more like mobile computers in their available uses. This has introduced additional difficulties for law enforcement officials in distinguishing one usage from another as drivers use their devices. This is more apparent in those countries which ban both handheld and hands-free usage, rather than those who have banned handheld use only, as officials cannot easily tell which function of the mobile phone is being used simply by looking at the driver. This can lead to drivers being stopped for using their device illegally on a phone call when, in fact, they were using the device for a legal purpose such as the phone's incorporated controls for car stereo or satnav.

A 2010 study reviewed the incidence of mobile phone use while cycling and its effects on behaviour and safety.[35]

Mobile banking and payments

Mobile payment system

Main articles: Mobile banking and Mobile payment
See also: Branchless banking and Contactless payment

In many countries, mobile phones are used to provide mobile banking services, which may include the ability to transfer cash payments by secure SMS text message. Kenya's M-PESA mobile banking service, for example, allows customers of the mobile phone operator Safaricom to hold cash balances which are recorded on their SIM cards. Cash may be deposited or withdrawn from M-PESA accounts at Safaricom retail outlets located throughout the country, and may be transferred electronically from person to person as well as used to pay bills to companies.

Branchless banking has also been successful in South Africa and the Philippines. A pilot project in Bali was launched in 2011 by the International Finance Corporation and an Indonesian bank Bank Mandiri.[36]

Another application of mobile banking technology is Zidisha, a US-based nonprofit micro-lending platform that allows residents of developing countries to raise small business loans from Web users worldwide. Zidisha uses mobile banking for loan disbursements and repayments, transferring funds from lenders in the United States to the borrowers in rural Africa using the Internet and mobile phones.[37]

Mobile payments were first trialled in Finland in 1998 when two Coca-Cola vending machines in Espoo were enabled to work with SMS payments. Eventually, the idea spread and in 1999 the Philippines launched the country's first commercial mobile payments systems on the mobile operators Globe and Smart.

Some mobile phone can make mobile payments via direct mobile billing schemes or through contactless payments if

the phone and point of sale support near field communication (NFC).[38] This requires the co-operation of manufacturers, network operators and retail merchants to enable contactless payments through NFC-equipped mobile phones.[39][40][41]

Tracking and privacy

See also: Cellphone surveillance and Mobile phone tracking

Mobile phones are also commonly used to collect location data. While the phone is turned on, the geographical location of a mobile phone can be determined easily (whether it is being used or not), using a technique known as multilateration to calculate the differences in time for a signal to travel from the cell phone to each of several cell towers near the owner of the phone.[42][43]

The movements of a mobile phone user can be tracked by their service provider and, if desired, by law enforcement agencies and their government. Both the SIM card and the handset can be tracked.[42]

China has proposed using this technology to track commuting patterns of Beijing city residents.[44] In the UK and US, law enforcement and intelligence services use mobiles to perform surveillance. They possess technology to activate the microphones in cell phones remotely in order to listen to conversations that take place near the phone.[45][46]

Thefts

According to the Federal Communications Commission, one out of three robberies involved the theft of a cellular phone. Police data in San Francisco showed that one-half of all robberies in 2012 were thefts of cellular phones. An online petition on Change.org called *Secure our Smartphones* urged smartphone manufacturers to install kill switches in their devices to make them unusable in case of theft. The petition is part of a joint effort by New York Attorney General Eric Schneiderman and San Francisco District Attorney George Gascón and was directed to the CEOs of the major smartphone manufacturers and telecommunication carriers.[47]

On Monday, 10 June 2013, Apple announced it would install a kill switch on its next iPhone operating system, due to debut in October 2013.[48]

1.1.6 Health effects

Main article: Mobile phone radiation and health
Further information: Mobile phones on aircraft

The effect mobile phone radiation has on human health is the subject of recent interest and study, as a result of the enormous increase in mobile phone usage throughout the world. Mobile phones use electromagnetic radiation in the microwave range, which some believe may be harmful to human health. A large body of research exists, both epidemiological and experimental, in non-human animals and in humans, of which the majority shows no definite causative relationship between exposure to mobile phones and harmful biological effects in humans. This is often paraphrased simply as the balance of evidence showing no harm to humans from mobile phones, although a significant number of individual studies do suggest such a relationship, or are inconclusive. Other digital wireless systems, such as data communication networks, produce similar radiation.

On 31 May 2011, the World Health Organization stated that mobile phone use may possibly represent a long-term health risk,[49][50] classifying mobile phone radiation as "possibly carcinogenic to humans" after a team of scientists reviewed studies on cell phone safety.[51] Mobile phones are in category 2B, which ranks it alongside coffee and other possibly carcinogenic substances.[52][53]

At least some recent studies have found an association between cell phone use and certain kinds of brain and salivary gland tumors. Lennart Hardell and other authors of a 2009 meta-analysis of 11 studies from peer-reviewed journals concluded that cell phone usage for at least ten years "approximately doubles the risk of being diagnosed with a brain tumor on the same ('ipsilateral') side of the head as that preferred for cell phone use."[54]

One study of past cell phone use cited in the report showed a "40% increased risk for gliomas (brain cancer) in the highest category of heavy users (reported average: 30 minutes per day over a 10-year period)."[55] This is a reversal from their prior position that cancer was unlikely to be caused by cellular phones or their base stations and that reviews had found no convincing evidence for other health effects.[50][56] Certain countries, including France, have warned against the use of cell phones especially by minors due to health risk uncertainties.[57] However, a study published 24 March 2012 in the *British Medical Journal* questioned these estimates, because the increase in brain cancers has not paralleled the increase in mobile phone use.[58]

1.1.7   Future evolution

Main article: 5G

5G is a technology used in research papers and projects to denote the next major phase of mobile telecommunication

standards beyond the 4G/IMT-Advanced standards. 5G is not officially used for any specification or official document yet made public by telecommunication companies or standardization bodies such as 3GPP, WiMAX Forum, or ITU-R. New standard releases beyond 4G are in progress by standardization bodies, but are at this time not considered as new mobile generations but under the 4G umbrella.

Deloitte is predicting a collapse in wireless performance to come as soon as 2016, as more devices using more and more services compete for limited bandwidth.[59]

1.1.8   Environmental impact

A mobile phone repair kiosk in Hong Kong

See also: Mobile phone recycling

Studies have shown that around 40-50% of the environmental impact of a mobile phone occurs during the manufacturing of the printed wiring boards and integrated circuits.[60] The average user replaces their mobile phone every 11 to 18 months.[61] The discarded phones then contribute to electronic waste.

Mobile phone manufacturers within Europe are subject to the WEEE directive. Australia introduced a mobile phone recycling scheme.[62]

1.1.9   Conflict minerals

See also: Conflict minerals

Demand for metals found in mobile phones and other electroncs fuelled the Second Congo War. The war claimed almost 5.5 million lives.[63] In a 2012 news story, *The Guardian* reported, "In unsafe mines deep underground in eastern Congo, children are working to extract minerals essential for the electronics industry. The profits from the minerals finance the bloodiest conflict since the second world war; the war has lasted nearly 20 years and has recently flared up again. ... For the last 15 years, the Democratic Republic of the Congo has been a major source of natural resources for the mobile phone industry."[64]

FairPhone is an attempt to develop a mobile phone which does not contain conflict minerals.

1.1.10 See also

1.1.11 References

[1] John F. Mitchell Biography

[2] Who invented the cell phone?

[3] Heeks, Richard (2008). "Meet Marty Cooper – the inventor of the mobile phone". *BBC* **41** (6): 26–33. doi:10.1109/MC.2008.192.

[4] "Mobile penetration". 9 July 2010.

[5] Sara Nagi. "Top 10 Best-selling Mobile Phone Brands in the World 2014". *TopTeny 2015*.

[6] Arthur C. Clarke: *Profiles of the Future* (1962, rev. eds. 1973, 1983, and 1999, Millennium edition with a new preface)

[7] "Swedish National Museum of Science and Technology". Tekniskamuseet.se. Retrieved 29 July 2009.

[8] UMTS World. "History of UMTS and 3G development". Umtsworld.com. Retrieved 29 July 2009.

[9] Fahd Ahmad Saeed. "Capacity Limit Problem in 3G Networks". Purdue School of Engineering. Retrieved 23 April 2010.

[10] Jeff Hecht. "Why Mobile Voice Quality Still Stinks—and How to Fix It". *ieee.org*.

[11] Elena Malykhina. "Why Is Cell Phone Call Quality So Terrible?". *scientificamerican.com*.

[12] Alan Henry. "What's the Best Mobile VoIP App?". *Lifehacker*. Gawker Media.

[13]

[14] "Nokia boosted by sales of cheap handsets". 20 October 2011.

[15] "Kosher Phones For Britain's Orthodox Jews". *Public Radio International*.

[16] "Introducing: A 'Kosher Phone' Permitted on Shabbat". *Arutz Sheva*.

[17] Tania Branigan (11 January 2010). "State owned China Mobile is world's biggest mobile phone operator". Guardian News and Media Limited. Retrieved 17 December 2011.

[18] Source: wireless intelligence

[19]

[20] "Samsung May Have Just Become The King Of Mobile Handsets, While S&P Downgrades Nokia To Junk". Retrieved 28 April 2012.

[21] "Gartner Says Sales of Smartphones Grew 20 Percent in Third Quarter of 2014". Gartner.

[22] "Annual Smartphone Sales Surpassed Sales of Feature Phones for the First Time in 2013". Retrieved 13 February 2014.

[23] "Worldwide Smartphone Shipments Top One Billion Units for the First Time". Retrieved 27 January 2014.

[24] "UK | Millions keep secret mobile". BBC News. 16 October 2001. Retrieved 4 November 2009.

[25] Brooks, Richard (13 August 2007). "Donated cell phones help battered women | San Bernardino County | PE.com | Southern California News | News for Inland Southern California". The Press-Enterprise. Retrieved 4 November 2009.

[26] Goodyear, Dana (7 January 2009). "Letter from Japan: I ♥ Novels". The New Yorker. Retrieved 29 July 2009.

[27] "You Witness News". News.yahoo.com. 26 January 2009. Retrieved 29 July 2009.

[28] Lynn, Jonathan. "Mobile phones help lift poor out of poverty: U.N. study". *Reuters*. Retrieved 2013-12-03.

[29] "4 Ways Smartphones Can Save Live TV". Tvgenius.net. Retrieved 4 June 2012.

[30] Donner, Jonathan, and Steenson, Molly Wright. "Beyond the Personal and Private: Modes of Mobile Phone Sharing in Urban India." In *The Reconstruction of Space and Time: Mobile Communication Practices*, edited by Scott Campbell and Rich Ling, 231–250. Piscataway, NJ: Transaction Publishers, 2008.

[31] Hahn, Hans and Kibora, Ludovic. "The Domestication of the Mobile Phone: Oral Society and New ICT in Burkina Faso". Journal of Modern African Studes 46 (2008): 87–109.

[32] http://www.itu.int/en/ITU-D/Statistics/Pages/stat/default.aspx

[33] source Informa 2007

[34] "Downloads_Guide". Netsize. Retrieved 29 July 2009.

[35] de Waard, D., Schepers, P., Ormel, W. and Brookhuis, K., 2010, *Mobile phone use while cycling: Incidence and effects on behaviour and safety, Ergonomics*, Vol 53, No. 1, January 2010, pp 30–42.

[36] "Branchless banking to start in Bali". The Jakarta Post. 13 April 2012. Retrieved 4 June 2012.

[37] ""Zidisha Set to "Expand" in Peer-to-Peer Microfinance", Microfinance Focus, Feb 2010". Microfinancefocus.com. 7 February 2010. Retrieved 4 June 2012.

[38] Feig, Nancy (25 June 2007). "Mobile Payments: Look to Korea". banktech.com. Retrieved 29 January 2011.

[39] Poulter, Sean (27 January 2011). "End of the credit card? With one swipe of an iPhone you'll be able to pay for your shopping". London: dailymail.co.uk. Retrieved 29 January 2011.

[40] Ready, Sarah (10 November 2009). "NFC mobile phone set to explode". connectedplanetonline.com. Retrieved 29 January 2011.

[41] Tofel, Kevin C. (20 August 2010). "VISA Testing NFC Memory Cards for Wireless Payments". gigaom.com. Retrieved 21 January 2011.

[42] "Tracking a suspect by mobile phone". *BBC News*. 3 August 2005. Retrieved 14 March 2009.

[43] Miller, Joshua (14 March 2009). "Cell Phone Tracking Can Locate Terrorists — But Only Where It's Legal". *FOX News*. Retrieved 4 February 2014.

[44] Cecilia Kang (3 March 2011). "China plans to track cellphone users, sparking human rights concerns". The Washington Post.

[45] McCullagh, Declan; Anne Broache (1 December 2006). "FBI taps cell phone mic as eavesdropping tool". *CNet News*. Retrieved 14 March 2009.

[46] Odell, Mark (1 August 2005). "Use of mobile helped police keep tabs on suspect". *Financial Times*. Retrieved 14 March 2009.

[47] Adams, Mike "Plea Urges Anti-Theft Phone Tech" *San Francisco Examiner* Friday, 7 June 2013 Page 5

[48] "Apple to add kill switches to help combat iPhone theft" by Jaxon Van Derbeken *San Francisco Chronicle* Tuesday, 11 June 2013 Page 1

[49] "IARC CLASSIFIES RADIOFREQUENCY ELECTROMAGNETIC FIELDS AS POSSIBLY CARCINOGENIC TO HUMANS" (PDF). *World Health Organization*.

[50] "What are the health risks associated with mobile phones and their base stations?". *Online Q&A*. World Health Organization. 5 December 2005. Retrieved 19 January 2008.

[51] "WHO: Cell phone use can increase possible cancer risk". CNN. 31 May 2011. Retrieved 31 May 2011.

[52] "Agents Classified by the IARC Monographs, Volumes 1–107" (PDF). *monographs.iarc.fr*.

[53] Kovvali, Gopala (1 January 2011). "Cell phones are as carcinogenic as coffee". *Journal of Carcinogenesis* **10** (1): 18. doi:10.4103/1477-3163.83044.

[54] Khurana, VG; Teo C; Kundi M; Hardell L; Carlberg M (2009). "Cell phones and brain tumors: A review including the long term epidemiologic data". *Surgical Neurology* **72** (3): 205–214. doi:10.1016/j.surneu.2009.01.019. PMID 19328536.

[55] "World Health Organization: Cell Phones May Cause Cancer". Business Insider. Retrieved 31 May 2011.

[56] "Electromagnetic fields and public health: mobile telephones and their base stations". *Fact sheet N°193*. World Health Organization. June 2000. Retrieved 19 January 2008.

[57] Brian Rohan (2 January 2008). "France warns against excessive mobile phone use". Reuters. Retrieved 10 May 2010.

[58] Little MP, Rajaraman P, Curtis RE, et al. (2012). "Mobile phone use and glioma risk: comparison of epidemiological study results with incidence trends in the United States". *BMJ* **344**: e1147. doi:10.1136/bmj.e1147.

[59] "Wireless performance will collapse, prices rise: Deloitte". *theregister.co.uk*.

[60] "The Secret Life Series - Environmental Impacts of Cell Phones". Inform, Inc. Retrieved 4 February 2014.

[61] "E-waste research group, Facts and figures". Griffith University. Retrieved 3 December 2011.

[62] "Mobile Phone Waste and The Environment". Aussie Recycling Program. Retrieved 3 December 2011.

[63] "Is your mobile phone helping fund war in Congo?". *The Daily Telegraph*. 27 September 2011.

[64] "Children of the Congo who risk their lives to supply our mobile phones". *The Guardian*. 7 December 2012.

1.1.12 Further reading

- Agar, Jon, *Constant Touch: A Global History of the Mobile Phone*, 2004 ISBN 1-84046-541-7
- Ahonen, Tomi, *m-Profits: Making Money with 3G Services*, 2002, ISBN 0-470-84775-1
- Ahonen, Kasper and Melkko, *3G Marketing* 2004, ISBN 0-470-85100-7
- Fessenden, R. A. (1908). "Wireless Telephony". *Annual Report of the Board of Regents of the Smithsonian Institution*: 161–196. Retrieved 7 August 2009.
- Glotz, Peter & Bertsch, Stefan, eds. *Thumb Culture: The Meaning of Mobile Phones for Society*, 2005
- Goggin, Gerard, *Global Mobile Media* (New York: Routledge, 2011), p. 176. ISBN 978-0415469180

- Jain, S. Lochlann. "Urban Errands: The Means of Mobility". Journal of Consumer Culture 2:3 (November 2002) 385–404. doi:10.1177/146954050200200305.

- Katz, James E. & Aakhus, Mark, eds. *Perpetual Contact: Mobile Communication, Private Talk, Public Performance*, 2002

- Kavoori, Anandam & Arceneaux, Noah, eds. *The Cell Phone Reader: Essays in Social Transformation*, 2006

- Kennedy, Pagan. Who Made That Cellphone?, *The New York Times*, 15 March 2013, p. MM19

- Kopomaa, Timo. *The City in Your Pocket*, Gaudeamus 2000

- Levinson, Paul, *Cellphone: The Story of the World's Most Mobile Medium, and How It Has Transformed Everything!*, 2004 ISBN 1-4039-6041-0

- Ling, Rich, *The Mobile Connection: the Cell Phone's Impact on Society*, 2004 ISBN 1-55860-936-9

- Ling, Rich and Pedersen, Per, eds. *Mobile Communications: Re-negotiation of the Social Sphere*, 2005 ISBN 1-85233-931-4

- Home page of Rich Ling

- Nyíri, Kristóf, ed. *Mobile Communication: Essays on Cognition and Community*, 2003

- Nyíri, Kristóf, ed. *Mobile Learning: Essays on Philosophy, Psychology and Education*, 2003

- Nyíri, Kristóf, ed. *Mobile Democracy: Essays on Society, Self and Politics*, 2003

- Nyíri, Kristóf, ed. *A Sense of Place: The Global and the Local in Mobile Communication*, 2005

- Nyíri, Kristóf, ed. *Mobile Understanding: The Epistemology of Ubiquitous Communication*, 2006

- Plant, Dr. Sadie, *on the mobile – the effects of mobile telephones on social and individual life*, 2001

- Rheingold, Howard, *Smart Mobs: The Next Social Revolution*, 2002 ISBN 0-7382-0861-2

- Singh, Rohit (April 2009). *Mobile phones for development and profit: a win-win scenario* (PDF). Overseas Development Institute. p. 2.

1.1.13 External links

- Media related to Mobile phones at Wikimedia Commons

- How Cell Phones Work at HowStuffWorks

- "The Long Odyssey of the Cell Phone", 15 photos with captions from *Time* magazine

- *Cell Phone, the ring heard around the world*—a video documentary by the Canadian Broadcasting Corporation

1.2 Mobile phone radiation and health

A man speaking on a mobile telephone

The effect of mobile phone radiation on human health is a subject of interest and study worldwide, as a result of the enormous increase in mobile phone usage throughout the world. As of November 2011, there were more than 6 billion subscriptions worldwide.[1] Mobile phones use electromagnetic radiation in the microwave range. Other digital wireless systems, such as data communication networks, produce similar radiation.

In 2011, International Agency for Research on Cancer (IARC) classified mobile phone radiation as Group 2B - possibly carcinogenic (*not* Group 2A - probably carcinogenic - nor the dangerous Group 1). That means that there "could be some risk" of carcinogenicity, so additional research into the long-term, heavy use of mobile phones needs to be conducted.[2] The WHO added in June 2011 that "to date, no adverse health effects have been established as being caused by mobile phone use",[3] a point they reiterated in October 2014.[4] Some national radiation advisory authorities[5] have recommended measures to minimize exposure to their citizens as a precautionary approach.

1.2.1 Effects

Many scientific studies have investigated possible health symptoms of mobile phone radiation. These studies are occasionally reviewed by some scientific committees to assess overall risks. A 2007 assessment published by the European Commission Scientific Committee on Emerging and Newly Identified Health Risks (SCENIHR)[6] concludes that the three lines of evidence, *viz.* animal, *in vitro*, and epidemiological studies, indicate that "exposure to RF fields is unlikely to lead to an increase in cancer in humans".

1.2.2 Radiation absorption

Part of the radio waves emitted by a mobile telephone handset are absorbed by the body. The radio waves emitted by a GSM handset are typically below a watt.[7] The maximum power output from a mobile phone is regulated by the mobile phone standard and by the regulatory agencies in each country. In most systems the cellphone and the base station check reception quality and signal strength and the power level is increased or decreased automatically, within a certain span, to accommodate different situations, such as inside or outside of buildings and vehicles.[8] The rate at which energy is absorbed by the human body is measured by the Specific Absorption Rate (SAR), and its maximum levels for modern handsets have been set by governmental regulating agencies in many countries. In the USA, the Federal Communications Commission (FCC) has set a SAR limit of 1.6 W/kg, averaged over a volume of 1 gram of tissue, for the head. In Europe, the limit is 2 W/kg, averaged over a volume of 10 grams of tissue. SAR values are heavily dependent on the size of the averaging volume. Without information about the averaging volume used, comparisons between different measurements cannot be made. Thus, the European 10-gram ratings should be compared among themselves, and the American 1-gram ratings should only be compared among themselves. SAR data for specific mobile phones, along with other useful information, can be found directly on manufacturers' websites, as well as on third party web sites.[9] It is worth noting that thermal radiation is not comparable to ionizing radiation in that it only increases the temperature in normal matter, it does not break molecular bonds or release electrons from their atoms.

Thermal effects

One well-understood effect of microwave radiation is dielectric heating, in which any dielectric material (such as living tissue) is heated by rotations of polar molecules induced by the electromagnetic field. In the case of a person using a cell phone, most of the heating effect will occur at the surface of the head, causing its temperature to increase by a fraction of a degree. In this case, the level of temperature increase is an order of magnitude less than that obtained during the exposure of the head to direct sunlight. The brain's blood circulation is capable of disposing of excess heat by increasing local blood flow. However, the cornea of the eye does not have this temperature regulation mechanism and exposure of 2–3 hours duration has been reported to produce cataracts in rabbits' eyes at SAR values from 100–140 W/kg, which produced lenticular temperatures of 41 °C. There were no cataracts detected in the eyes of monkeys exposed under similar conditions.[10] Premature cataracts have not been linked with cell phone use, possibly because of the lower power output of mobile phones.

Non-thermal effects

The communications protocols used by mobile phones often result in low-frequency pulsing of the carrier signal. While the existence of effects due to the field is undisputable, whether these modulations are causing these effects or these are still of thermic nature is subject to debate.[11]

Some researchers have argued that so-called "non-thermal effects" could be reinterpreted as a normal cellular response to an increase in temperature. The German biophysicist Roland Glaser, for example,[12] has argued that there are several thermoreceptor molecules in cells, and that they activate a cascade of second and third messenger systems, gene expression mechanisms and production of heat shock proteins in order to defend the cell against metabolic cell stress caused by heat. The increases in temperature that cause these changes are too small to be detected by studies such as REFLEX, which base their whole argument on the apparent stability of thermal equilibrium in their cell cultures.

Other researchers believe the stress proteins are unrelated to thermal effects, since they occur for both extremely low frequencies (ELF) and radio frequencies (RF), which have very different energy levels.[13] Another preliminary study published in 2011 by *The Journal of the American Medical Association* conducted using fluorodeoxyglucose injections and positron emission tomography concluded that exposure to radiofrequency signal waves within parts of the brain closest to the cell phone antenna resulted in increased levels of glucose metabolism, but the clinical significance of this finding is unknown.[14][15]

Blood–brain barrier effects Swedish researchers from Lund University (Salford, Brun, Persson, Eberhardt, and Malmgren) have studied the effects of microwave radiation on the rat brain. They found a leakage of albumin into the brain via a permeated blood–brain barrier.[16][17] This confirms earlier work on the blood–brain barrier by Allan Frey, Oscar and Hawkins, and Albert and Kerns.[18] Other groups have not confirmed these findings in vitro cell studies[19] or whole animal studies.[18]:102[20]

Prof Leszczynski of Finland's radiation and nuclear safety authority found that, at the maximum legal limit for mobile radiation, one protein in particular, HSP 27, was affected. HSP 27 played a critical role in the integrity of the blood-brain barrier.[21]

Cancer In 2006, a large Danish group's study about the connection between mobile phone use and cancer incidence was published. It followed over 420,000 Danish citizens for 20 years and showed no increased risk of cancer.[22] A 2011 follow-up confirmed these findings.[23]

The following studies of long time exposure have been published:

- The 13 nation INTERPHONE project – the largest study of its kind ever undertaken – was published in 2010 and did not find a solid link between mobile phones and brain tumours.[24]

The *International Journal of Epidemiology* published[25] a combined data analysis from INTERPHONE, a multi national population-based case-control study of glioma and meningioma, the most common types of brain tumour.

The authors reported the following conclusion:

> Overall, no increase in risk of glioma or meningioma was observed with use of mobile phones. There were suggestions of an increased risk of glioma at the highest exposure levels, but biases and error prevent a causal interpretation. The possible effects of long-term heavy use of mobile phones require further investigation.

In the press release[26] accompanying the release of the paper, Dr. Christopher Wild, Director of the International Agency for Research on Cancer (IARC) said:

> An increased risk of brain cancer is not established from the data from Interphone. However, observations at the highest level of cumulative call time and the changing patterns of mobile phone use since the period studied by Interphone,

particularly in young people, mean that further investigation of mobile phone use and brain cancer risk is merited.

A number of independent health and government authorities have commented on this important study including The Australian Centre for Radiofrequency Bioeffects Research (ACRBR) which said in a statement that:[27]

> Until now there have been concerns that mobile phones were causing increases in brain tumours. Interphone is both large and rigorous enough to address this claim, and it has not provided any convincing scientific evidence of an association between mobile phone use and the development of glioma or meningioma. While the study demonstrates some weak evidence of an association with the highest tenth of cumulative call time (but only in those who started mobile phone use most recently), the authors conclude that biases and errors limit the strength of any conclusions in this group. It now seems clear that if there was an effect of mobile phone use on brain tumour risks in adults, this is likely to be too small to be detectable by even a large multinational study of the size of Interphone.

The Australian Radiation Protection and Nuclear Safety Agency (ARPANSA) which said in a statement that:

> On the basis of current understanding of the relationship between brain cancer and use of mobile phones, including the recently published data from the INTERPHONE study, ARPANSA:
>
> **concludes** that currently available data do not warrant any general recommendation to limit use of mobile phones in the adult population,
>
> **continues** to inform those concerned about potential health effects that they may limit their exposure by reducing call time, by making calls where reception is good, by using hands-free devices or speaker options, or by texting; and
>
> **recommends** that, due to the lack of any data relating to children and long term use of mobile phones, parents encourage their children to limit their exposure by reducing call time, by making calls where reception is good, by using hands-free devices or speaker options, or by texting.

The Cancer Council Australia said in a statement that it cautiously welcomed the results of the largest international study to date into mobile phone use, which has found no evidence that normal use of mobile phones, for a period up to 12 years, can cause brain cancer.

Chief Executive Officer, Professor Ian Olver, said findings from the Interphone study, conducted across 13 countries including Australia, were consistent with other research that had failed to find a link between mobile phones and cancer.

> This supports previous research showing mobile phones don't damage cell DNA, meaning they can't cause the type of genetic mutations that develop into cancer," Professor Olver said.

> However, it has been suggested that electromagnetic fields associated with mobile phones may play a role in speeding up the development of an existing cancer. The Interphone study found no evidence to support this theory.

- A Danish study (2004) that took place over 10 years found no evidence to support a link. However, this study has been criticized for collecting data from subscriptions and not necessarily from actual users. It is known that some subscribers do not use the phones themselves but provide them for family members to use. That this happens is supported by the observation that only 61% of a small sample of the subscribers reported use of mobile phones when responding to a questionnaire.[22][28]

- A Swedish study (2005) that draws the conclusion that "the data do not support the hypothesis that mobile phone use is related to an increased risk of glioma or meningioma."[29]

- A British study (2005) that draws the conclusion that "The study suggests that there is no substantial risk of acoustic neuroma in the first decade after starting mobile phone use. However, an increase in risk after longer term use or after a longer lag period could not be ruled out."[30]

- A German study (2006) that states "In conclusion, no overall increased risk of glioma or meningioma was observed among these cellular phone users; however, for long-term cellular phone users, results need to be confirmed before firm conclusions can be drawn."[31]

- A joint study conducted in northern Europe that draws the conclusion that "Although our results overall do not indicate an increased risk of glioma in relation to mobile phone use, the possible risk in the most heavily exposed part of the brain with long-term use needs to be explored further before firm conclusions can be drawn."[32]

Other studies on cancer and mobile phones are:

- A Swedish scientific team at the Karolinska Institute conducted an epidemiological study (2004) that suggested that regular use of a mobile phone over a decade or more was associated with an increased risk of acoustic neuroma, a type of benign brain tumor. The increase was not noted in those who had used phones for fewer than 10 years.[33]

- The INTERPHONE study group from Japan published the results of a study of brain tumour risk and mobile phone use. They used a new approach: determining the SAR inside a tumour by calculating the radio frequency field absorption in the exact tumour location. Cases examined included glioma, meningioma, and pituitary adenoma. They reported that the overall odds ratio (OR) was not increased and that there was no significant trend towards an increasing OR in relation to exposure, as measured by SAR.[34]

In 2007, Dr. Lennart Hardell, from Örebro University in Sweden, reviewed published epidemiological papers (2 cohort studies and 16 case-control studies) and found that:[35]

- Cell phone users had an increased risk of malignant gliomas.

- Cell phone use was linked to a higher rate of acoustic neuromas.

- Tumors are more likely to occur on the side of the head that the cell handset is used.

- One hour of cell phone use per day significantly increases tumor risk after ten years or more.

In a February 2008 update on the status of the INTERPHONE study IARC stated that the long-term findings '...could either be causal or artifactual, related to differential recall between cases and controls.'[36]

A publication titled "Public health implications of wireless technologies" cites that Lennart Hardell found age is a significant factor. The report repeated the finding that the use of cell phones before age 20 increased the risk of brain tumors by 5.2, compared to 1.4 for all ages.[37] A review by

Hardell et al. concluded that current mobile phones are not safe for long-term exposure.[38]

In a time trends study in Europe, conducted by the Institute of Cancer Epidemiology in Copenhagen, no significant increase in brain tumors among cell phone users was found between the years of 1998 and 2003. "The lack of a trend change in incidence from 1998 to 2003 suggests that the induction period relating mobile phone use to brain tumors exceeds 5–10 years, the increased risk in this population is too small to be observed, the increased risk is restricted to subgroups of brain tumors or mobile phone users, or there is no increased risk."[39]

On 31 May 2011, the International Agency for Research on Cancer classified radiofrequency electromagnetic fields as possibly carcinogenic to humans (Group 2B). The IARC assessed and evaluated available literature and studies about the carcinogenicity of radiofrequency electromagnetic fields (RF-EMF), and found the evidence to be "limited for carcinogenicity of RF-EMF, based on positive associations between glioma and acoustic neuroma and exposure". The conclusion of the IARC was mainly based on the INTERPHONE study, which found an increased risk for glioma in the highest category of heavy users (30 minutes per day over a 10-year period), although no increased risk was found at lower exposure and other studies could not back up the findings. The evidence for other types of cancer was found to be "inadequate". Some members of the Working Group opposed the conclusions and considered the current evidence in humans still as "inadequate", citing inconsistencies between the assessed studies.[2][40]

In 2011, a review published in *Environmental Health Perspectives* found that increasing evidence suggests that mobile phone use does not cause brain tumors in adults.[41]

In 2012, a systematic review was published which found "no statistically significant increase in risk for adult brain cancer or other head tumors from wireless phone use."[42]

Researchers at the National Cancer Institute found that while cell phone use increased substantially over the period 1992 to 2008 (from nearly zero to almost 100 percent of the population), the U.S. trends in glioma incidence did not mirror that increase.[43]

In 2014, a French national case-control study, CERENAT, confirmed a possible association between heavy mobile phone use and brain tumours (gliomas and meningiomas), finding an up to eight-fold increased risk of gliomas tied with cellphone use.[44]

In March 2015, a study on mice carried on by Prof. Alexander Lerchl of Jacobs University in Bremen and his team on behalf of the German Federal Office for Radiation Protection found out that the growth rate of liver and lung cancer generated by chemical substances raises substantially when the animals are lifelong irradiated with mobile phone-like e.m. fields.[45] This study confirms a research carried on in 2010 at Fraunhofer Institute.[45] Moreover, the researchers discovered a significant higher rate of lymphomas, and found out that some of the effects occur also for field intensities lower than current limits.[45] The underlying mechanisms are unknown.[45]

Systematic reviews and meta-analyses   In 2009, a meta-analysis of 23 studies on mobile phone use and tumor risk found that "there is possible evidence" that mobile phone use causes an increased risk of tumors.[46]

Cognitive effects   A 2009 study, examined the effects of exposure to radiofrequency radiation (RFR) emitted by standard GSM cell phones on the cognitive functions of humans. The study confirmed longer (slower) response times to a spatial working memory task when exposed to RFR from a standard GSM cellular phone placed next to the head of male subjects, and showed that longer duration of exposure to RFR may increase the effects on performance. Right-handed subjects exposed to RFR on the left side of their head on average had significantly longer response times when compared to exposure to the right side and sham-exposure.[47]

Electromagnetic hypersensitivity   Main article: Electromagnetic hypersensitivity

Some users of mobile handsets have reported feeling several unspecific symptoms during and after its use; ranging from burning and tingling sensations in the skin of the head and extremities, fatigue, sleep disturbances, dizziness, loss of mental attention, reaction times and memory retentiveness, headaches, malaise, tachycardia (heart palpitations), to disturbances of the digestive system. Reports have noted that all of these symptoms can also be attributed to stress and that current research cannot separate the symptoms from nocebo effects.[48]

Genotoxic effects   A meta-analysis (2008) of 63 in vitro and in vivo studies from the years 1990–2005, concluded that RF radiation was genotoxic only in some conditions and that the studies reporting positive effects evidenced publication bias.[49]

A meta-study (2009) of 101 publications on genotoxicity of RF electromagnetic fields, showed that 49 reported a genotoxic effect and 42 not. The authors found "ample evidence that RF-EMF can alter the genetic material of exposed cells in vivo and in vitro and in more than one way".[50]

In 1995, in the journal *Bioelectromagnetics*, Henry Lai and Narenda P. Singh reported damaged DNA after two hours of microwave radiation at levels deemed safe according to U.S. government standards.[51]

In December 2004, a pan-European study named REFLEX (Risk Evaluation of Potential Environmental Hazards from Low Energy Electromagnetic Field (EMF) Exposure Using Sensitive in vitro Methods), involving 12 collaborating laboratories in several countries showed some compelling evidence of DNA damage of cells in in-vitro cultures, when exposed between 0.3 to 2 watts/kg, whole-sample average. There were indications, but not rigorous evidence of other cell changes, including damage to chromosomes, alterations in the activity of certain genes and a boosted rate of cell division.[52]

Research published in 2004, by a team at the University of Athens, had a reduction in reproductive capacity in fruit flies exposed to 6 minutes of 900 MHz pulsed radiation for five days.[53]

Subsequent research, again conducted on fruit flies, was published in 2007, with the same exposure pattern but conducted at both 900 MHz and 1800 MHz, and had similar changes in reproductive capacity with no significant difference between the two frequencies.[54]

Following additional tests published in a third article, the authors stated they thought their research suggested the changes were "...due to degeneration of large numbers of egg chambers after DNA fragmentation of their constituent cells ...".[55]

Australian research conducted in 2009, by subjecting in vitro samples of human spermatozoa to radio-frequency radiation at 1.8 GHz and specific absorption rates (SAR) of 0.4 to 27.5 W/kg showed a correlation between increasing SAR and decreased motility and vitality in sperm, increased oxidative stress and 8-Oxo-2'-deoxyguanosine markers, stimulating DNA base adduct formation and increased DNA fragmentation.[56]

Sleep and EEG effects Sleep, EEG and waking rCBF have been studied in relation to RF exposure for a decade now, and the majority of papers published to date have found some form of effect. While a Finnish study failed to find any effect on sleep or other cognitive function from pulsed RF exposure,[57] most other papers have found significant effects on sleep.[58][59][60][61][62][63] Two of these papers found the effect was only present when the exposure was pulsed (amplitude modulated), and one early paper found that sleep quality (measured by the amount of participants' broken sleep) improved.

While some papers were inconclusive or inconsistent,[64][65] a number of studies have now demonstrated reversible

EEG and rCBF alterations from exposure to pulsed RF exposure.[66][67][68][69] German research from 2006 found that statistically significant EEG changes could be consistently found, but only in a relatively low proportion of study participants (12 - 30%).[70]

Behavioural effects A study on mice offspring suggested that cell phone use during pregnancy may cause behavioural problems that resemble the effects of ADHD.[71]

Sperm count and sperm quality A number of studies have shown relationships between mobile telephone use and reduced sperm count and sperm quality. Peer reviewed studies have shown relationships using statistical questionnaire techniques,[72][73][74][75][76] controlled experiments on living humans,[77] and controlled experiments on sperm outside the body.[78][79][80][81]

The Environmental Working Group (EWG) has a web page entitled "Cell Phone Radiation Damages Sperm, Studies Show" published August 2013. The EWG page reviews and tabulates studies showing relationships between mobile phone use and low sperm count and sperm quality.[82]

1.2.3   Health hazards of base stations

A Greenfield-type tower used in base stations for mobile telephony

Another area of concern is the radiation emitted by the fixed infrastructure used in mobile telephony, such as base stations and their antennas, which provide the link to and from mobile phones. This is because, in contrast to mobile handsets, it is emitted continuously and is more powerful at close quarters. On the other hand, field intensities drop rapidly with distance away from the base of transmitters because of the attenuation of power with the square of distance.

One popular design of mobile phone antenna is the sector antenna, whose coverage is 120 degrees horizontally and about ∓5 degrees from the vertical.

Because base stations operate at less than 100 watts and the antenna is raised up well above ground, the radiation at ground level is much weaker than a cell phone due to the power relationship appropriate for that design of antenna. Base station emissions must comply with safety guidelines (see *Safety standards and licensing* below). Some countries, however (such as South Africa, for example), have no health regulations governing the placement of base stations.

Several surveys have found a variety of self-reported symptoms for people who live close to base stations.[83][84][85][86][87] However, there are significant challenges in conducting studies of populations near base stations, especially in assessment of individual exposure.[88] Self-report studies can also be vulnerable to the nocebo effect.

Two double-blind placebo-controlled trials conducted at the University of Essex and another in Switzerland[89] concluded that mobile phone masts were unlikely to be causing these short-term effects in a group of volunteers who complained of such symptoms.[90] The Essex study found that subjects were unable to tell whether they were being exposed to electromagnetic fields or not, and that sensitive subjects reported lower well-being independently of exposure. The principal investigator concluded "It is clear that sensitive individuals are suffering real symptoms and often have a poor quality of life. It is now important to determine what other factors could be causing these symptoms, so appropriate research studies and treatment strategies can be developed."

Experts consulted by France considered it was mandatory that the main antenna axis should not to be directly in front of a living place at a distance shorter than 100 metres.[91] This recommendation was modified in 2003[92] to say that antennas located within a 100-metre radius of primary schools or childcare facilities should be better integrated into the cityscape and was not included in a 2005 expert report.[93] The Agence française de sécurité sanitaire environnementale currently says that there is no demonstrated short-term effect of electromagnetic fields on health, but that there are open questions for long-term effects, and that it's easy to reduce exposure via technological improvements.[94]

1.2.4 Occupational health hazards

Telecommunication workers who spend time at a short distance from the active equipment, for the purposes of testing, maintenance, installation, etcetera, may be at risk of much greater exposure than the general population. Many times base stations are not turned off during maintenance, but the power being sent through to the antennas is cut off, so that the workers do not have to work near live antennas.

A variety of studies over the past 50 years have been done on workers exposed to high RF radiation levels; studies including radar laboratory workers, military radar workers, electrical workers, and amateur radio operators. Most of these studies found no increase in cancer rates over the general population or a control group. Many positive results could have been attributed to other work environment conditions, and many negative results (reduced cancer rates) also occurred.[95]

1.2.5 Safety standards and licensing

In order to protect the population living around base stations and users of mobile handsets, governments and regulatory bodies adopt safety standards, which translate to limits on exposure levels below a certain value. There are many proposed national and international standards, but that of the International Commission on Non-Ionizing Radiation Protection (ICNIRP) is the most respected one, and has been adopted so far by more than 80 countries. For radio stations, ICNIRP proposes two safety levels: one for occupational exposure, another one for the general population. Currently there are efforts underway to harmonise the different standards in existence.[96]

Radio base licensing procedures have been established in the majority of urban spaces regulated either at municipal/county, provincial/state or national level. Mobile telephone service providers are, in many regions, required to obtain construction licenses, provide certification of antenna emission levels and assure compliance to ICNIRP standards and/or to other environmental legislation.

Many governmental bodies also require that competing telecommunication companies try to achieve sharing of towers so as to decrease environmental and cosmetic impact. This issue is an influential factor of rejection of installation of new antennas and towers in communities.

The safety standards in the U.S. are set by the Federal Communications Commission (FCC). The FCC has based its standards primarily on those standards established by the Institute of Electrical and Electronics Engineers (IEEE), specifically Subcommittee 4 of the "International Committee on Electromagnetic Safety".

Switzerland has set safety limits lower than the ICNIRP limits for certain "sensitive areas" (classrooms, for example).[97]

1.2.6 Lawsuits

In the USA, a small number of personal injury lawsuits have been filed by individuals against cellphone manufacturers, such as Motorola,[98] NEC, Siemens and Nokia, on the basis of allegations of causation of brain cancer and death. In US federal court, expert testimony relating to science must be first evaluated by a judge, in a Daubert hearing, to be relevant and valid before it is admissible as evidence. In one case against Motorola, the plaintiffs alleged that the use of wireless handheld telephones could cause brain cancer, and that the use of Motorola phones caused one plaintiff's cancer. The judge ruled that no sufficiently reliable and relevant scientific evidence in support of either general or specific causation was proffered by the plaintiffs; accepted a motion to exclude the testimony of the plaintiffs' experts; and denied a motion to exclude the testimony of the defendants' experts.[99] This particular case is from 2002.

French High Court ruling against telecom company

In February 2009, the telecom company Bouygues Telecom was ordered to take down a mobile phone mast due to uncertainty about its effect on health. Residents in the commune Charbonnières in the Rhône department had sued the company claiming adverse health effects from the radiation emitted by the 19 meter tall antenna.[100] The milestone ruling by the Versailles Court of Appeal reversed the burden of proof which is usual in such cases by emphasizing the extreme divergence between different countries in assessing safe limits for such radiation. The court stated that, "Considering that, while the reality of the risk remains hypothetical, it becomes clear from reading the contributions and scientific publications produced in debate and the divergent legislative positions taken in various countries, that uncertainty over the harmlessness of exposure to the waves emitted by relay antennas persists and can be considered serious and reasonable".[101]

Italian High Court ruling in favour of "causal" link with brain cancer

In October 2012, Italian high court (Corte suprema di cassazione) granted an Italian businessman, *Innocente Marcoloni* a pension for occupational disease; "[c]ontrary to the denials of many health agencies in the U.S. and in some

other countries, the Italian Supreme Court has recognized a "causal" link between heavy mobile phone use and brain tumor risk in a worker's compensation case."[102][103] According to Reuters, a lower court in Brescia had "ruled there was a causal link between the use of mobile and cordless telephones and tumours" in the case of "Innocenzo Marcolini who developed a tumour in the left side of his head after using his mobile phone for [between 5 and 6] hours a day for 12 years. He normally held the phone in his left hand, while taking notes with his right hand" and that the ruling was upheld but they summarized experts saying the "decision flies in the face of much scientific opinion, which generally says there is not enough evidence to declare a link between mobile phone use and diseases such as cancer and some experts said the Italian ruling should not be used to draw wider conclusions about the subject."[104] As it takes time to develop cancer, the court disregarded short-term studies. The court based their ruling on "studies conducted between 2005 and 2009 by a group led by Lennart Hardell, a cancer specialist at the University Hospital in Orebro in Sweden"[104] and disregarded studies that were even partially funded by the mobile phone industry such as the INTERPHONE (see above).

Indian citizens against telecom company

In 2012, a case was filed against the mobile towers in residential areas, schools and hospitals.[105] In March 2013, based on the WHO notification dated 31 May 2011,[106] wherein the mobile tower radiations had been classified as possibly carcinogenic and the research conducted by the scientists of IIT Kharagpur, India,[107] a writ was filed by Advocate Vikas Nagwan for the suspected death of one Late. Hemant Sharma[108] for removal of the mobile towers from residential areas.

1.2.7 Precaution

Precautionary principle

In 2000, the World Health Organization (WHO) recommended that the precautionary principle could be voluntarily adopted in this case.[109] It follows the recommendations of the European Community for environmental risks. According to the WHO, the "precautionary principle" is "a risk management policy applied in circumstances with a high degree of scientific uncertainty, reflecting the need to take action for a potentially serious risk without awaiting the results of scientific research." Other less stringent recommended approaches are prudent avoidance principle and as low as reasonably practicable. Although all of these are problematic in application, due to the widespread use and economic importance of wireless telecommunication systems in modern civilization, there is an increased popularity of such measures in the general public, though also evidence that such approaches may increase concern.[110] They involve recommendations such as the minimization of cellphone usage, the limitation of use by at-risk population (such as children), the adoption of cellphones and microcells with as low as reasonably practicable levels of radiation, the wider use of hands-free and earphone technologies such as Bluetooth headsets, the adoption of maximal standards of exposure, RF field intensity and distance of base stations antennas from human habitations, and so forth. Overall, public information remains a challenge as various health consequences are evoked in the literature and by the media, putting populations under chronic exposure to potentially worrying information.[111]

Precautionary measures and health advisories

In May 2011, the World Health Organisation's International Agency for Research on Cancer announced it was classifying electromagnetic fields from mobile phones and other sources as "possibly carcinogenic to humans" and advised the public to adopt safety measures to reduce exposure, like use of hands-free devices or texting.

Some national radiation advisory authorities, including those of Austria,[5] France,[112] Germany,[113] and Sweden,[114] have recommended measures to minimize exposure to their citizens. Examples of the recommendations are:

- Use hands-free to decrease the radiation to the head.

- Keep the mobile phone away from the body.

- Do not use telephone in a car without an external antenna.

The use of "hands-free" was not recommended by the British Consumers' Association in a statement in November 2000 as they believed that exposure was increased.[115] However, measurements for the (then) UK Department of Trade and Industry[116] and others for the French l'Agence française de sécurité sanitaire environnementale[117] showed substantial reductions. In 2005 Professor Lawrie Challis and others said clipping a ferrite bead onto hands-free kits stops the radio waves travelling up the wire and into the head.[118]

Several nations have advised moderate use of mobile phones for children.[119] A journal by Gandhi et al. in 2006 states that children receive higher levels of SAR. When 5- and 10- year olds are compared to adults, they receive about 153% higher SAR levels. Also, with the permittivity of the brain decreasing as one gets older and the higher relative volume of the exposed growing brain in children, radiation penetrates far beyond the mid-brain.[120]

1.2.8 See also

- Wireless electronic devices and health

- Background radiation

- Bioelectromagnetism

- BioInitiative Report

- COSMOS cohort study

- Electromagnetic hypersensitivity

- Electromagnetic radiation and health

- Microwave News

- Mobile phones and driving safety

- Non-ionizing radiation

- Possible health effects of body scanners

- Radiation biology

1.2.9 References

[1] "Market Data Summary (Q2 2009)". GSM Association. Retrieved 30 January 2010.

[2] "IARC classifies radiofrequency electromagnetic fields as possibly carcinogenic to humans" (PDF). *World Health Organization press release N° 208* (Press release). International Agency for Research on Cancer. 31 May 2011. Retrieved 2 June 2011.

[3] "Electromagnetic fields and public health: mobile phones - Fact sheet N°193". World Health Organization. June 2011. Archived from the original on 2011-08-14. Retrieved 2015-05-22. (Use archive link for archived copy of June 2011 version of Fact sheet No 193).

[4] "Electromagnetic fields and public health: mobile phones - Fact sheet N°193". World Health Organization. October 2014. Retrieved 2015-05-22.

[5] "Information: Wie gefährlich sind Handystrahlen wirklich?" (in German). Marktgemeinde Pressbaum. Archived from the original on 2011-10-02. Retrieved 16 May 2015.

[6] "Conclusions on mobile phones and radio frequency fields". European Commission Scientific Committee on Emerging and Newly Identified Health Risks (SCENIHR). Retrieved 8 December 2008.

[7] "GSM phone signal analysis" (pdf).

[8] "Output Power-Control Loop Design for GSM Mobile Phones" (pdf). Retrieved 12 February 2013.

[9] For example, two listings using the European 10 g standard: of more current models at "Mobile Phones UK". *Mobile Phones UK web site*. Landmark Internet Ltd. Retrieved 19 January 2008.; of phones from 2005 and earlier at "The Complete SAR List For All Phones (Europe)". On-Line-Net - Web Design & Internet Services (as SARValues.com). Retrieved 19 January 2008. (a listing of US phones from 2005 and earlier, using the US 1 g standard, is also available at the SARValues site)

[10] International Commission on Non-Ionizing Radiation Protection (April 1998). "Guidelines For Limiting Exposure To Time-Varying Electric, Magnetic, And Electromagnetic Fields (up to 300 GHz)" (PDF). *Health Physics* **74** (4): 494–505. PMID 9525427. Retrieved 28 March 2010.

[11] Foster, Kenneth R.; Repacholi, Michael H. (2004). "Biological Effects of Radiofrequency Fields: Does Modulation Matter?". *Radiation Research* **162** (2): 219–25. doi:10.1667/RR3191. PMID 15387150.

[12] Glaser, Roland (December 2005). "Are thermoreceptors responsible for "non-thermal" effects of RF fields?" (PDF). *Edition Wissenschaft* (Bonn, Germany: Forschungsgemeinschaft Funk) (21). OCLC 179908725. Retrieved 19 January 2008.

[13] Blank, Martin; Goodman, Reba (2009). "Electromagnetic fields stress living cells". *Pathophysiology* **16** (2–3): 71–8. doi:10.1016/j.pathophys.2009.01.006. PMID 19268550.

[14] Volkow, Nora D.; Tomasi, Dardo; Wang, Gene-Jack; Vaska, Paul; Fowler, Joanna S.; Telang, Frank; Alexoff, Dave; Logan, Jean; et al. (2011). "Effects of Cell Phone Radiofrequency Signal Exposure on Brain Glucose Metabolism". *JAMA* **305** (8): 808–13. doi:10.1001/jama.2011.186. PMC 3184892. PMID 21343580.

[15] Press, Canadian (23 February 2011). "Cellphones linked to increased brain glucose metabolism". *The Globe and Mail* (Toronto). Retrieved 26 February 2011.

[16] Salford, Leif G.; Arne E. Brun; Jacob L. Eberhardt; Lars Malmgren; Bertil R. R. Persson (June 2003). "Nerve Cell Damage in Mammalian Brain after Exposure to Microwaves from GSM Mobile Phones". *Environmental Health Perspectives* (United States: National Institute of Environmental Health Sciences) **111** (7): 881–883. doi:10.1289/ehp.6039. PMC 1241519. PMID 12782486. Retrieved 8 January 2008.

[17] Salford, Leif G.; Henrietta Nittby; Arne Brun; Gustav Grafstrom; Lars Malmgren; Marianne Sommarin; Jacob Eberhardt; Bengt Widegren; Bertil R. R. Persson (2008). "The Mammalian Brain in the Electromagnetic Fields Designed by Man with Special Reference to Blood-Brain Barrier Function, Neuronal Damage and Possible Physical Mechanisms". *Progress of Theoretical Physics Supplement* (Japan: Physical Society of Japan) **173**: 283–309. doi:10.1143/PTPS.173.283.

[18] Frey, Allan (March 1998). "Headaches from Cellular Telephones: Are They Real and What Are the Implications?". *Environmental Health Perspectives* **106** (3): 101–3. doi:10.1289/ehp.98106101. PMC 1533043. PMID 9441959.

[19] Franke; et al. (2 September 2005). "Electromagnetic fields (GSM 1800) do not alter blood–brain barrier permeability to sucrose in models in vitro with high barrier tightness". *Bioelectromagnetics* **26** (7): 529–535. doi:10.1002/bem.20123.

[20] Kuribayashi; et al. "Lack of effects of 1439 MHz electromagnetic near field exposure on the blood–brain barrier in immature and young rats". *Bioelectromagnetics* **26** (7): 578–588. doi:10.1002/bem.20138.

[21] http://www.theguardian.com/uk/2002/jun/20/research.medicalscience

[22] Schüz, J; Jacobsen, R; Olsen, JH; Boice, JD; McLaughlin, JK; Johansen, C (December 2006). "Cellular Telephone Use and Cancer Risk: Update of a Nationwide Danish Cohort". *Journal of the National Cancer Institute* **98** (23): 1707–1713. doi:10.1093/jnci/djj464. PMID 17148772. Retrieved 20 January 2008. Among long-term subscribers of 10 years or more, cellular telephone use was not associated with increased risk for brain tumors ..., and there was no trend with time since first subscription. ...CONCLUSIONS: We found no evidence for an association between tumor risk and cellular telephone use among either short-term or long-term users. Moreover, the narrow confidence intervals provide evidence that any large association of risk of cancer and cellular telephone use can be excluded.

[23] "Use of mobile phones and risk of brain tumours: update of Danish cohort study", *BMJ* (343), 2011, doi:10.1136/bmj.d6387, PMC 3197791, PMID 22016439

[24] Parker-Pope, Tara (6 June 2011). "Piercing the Fog Around Cellphones and Cancer". *The New York Times*.

[25] Interphone Study Group (2010). "Brain tumour risk in relation to mobile telephone use: Results of the INTERPHONE international case-control study". *International Journal of Epidemiology* **39** (3): 675–694. doi:10.1093/ije/dyq079. PMID 20483835.

[26] "Interphone study reports on mobile phone use and brain cancer risk" (PDF) (Press release). International Agency for Research on Cancer. 17 May 2010. Retrieved 6 June 2011.

[27] http://www.acrbr.org.au/FAQ/ACRBR%20Interphone%20Position%20Statement%20May2010.pdf

[28] Ahlbom, Anders; Feychting, Maria; Cardis, Elisabeth; Elliott, Paul (2007). "Re: Cellular Telephone Use and Cancer Risk: Update of a Nationwide Danish Cohort Study". *Journal of the National Cancer Institute* **99** (8): 655–655. doi:10.1093/jnci/djk143. PMID 17440169.

[29] Lönn, Stefan; Ahlbom, Anders; Hall, Per; Feychting, Maria; Swedish Interphone Study Group (2005). "Long-Term Mobile Phone Use and Brain Tumor Risk". *American Journal of Epidemiology* **161** (6): 526–35. doi:10.1093/aje/kwi091. PMID 15746469.

[30] Schoemaker, M J; Swerdlow, A J; Ahlbom, A; Auvinen, A; Blaasaas, K G; Cardis, E; Christensen, H Collatz; Feychting, M; et al. (2005). "Mobile phone use and risk of acoustic neuroma: Results of the Interphone case–control study in five North European countries". *British Journal of Cancer* **93** (7): 842–848. doi:10.1038/sj.bjc.6602764. PMC 2361634. PMID 16136046.

[31] Schüz, Joachim; Böhler, Eva; Berg, Gabriele; Schlehofer, Brigitte; Hettinger, Iris; Schlaefer, Klaus; Wahrendorf, Jürgen; Kunna-Grass, Katharina; et al. (2006). "Cellular Phones, Cordless Phones, and the Risks of Glioma and Meningioma (Interphone Study Group, Germany)". *American Journal of Epidemiology* **163** (6): 512–20. doi:10.1093/aje/kwj068. PMID 16443797.

[32] Lahkola, Anna; Auvinen, Anssi; Raitanen, Jani; Schoemaker, Minouk J.; Christensen, Helle C.; Feychting, Maria; Johansen, Christoffer; Klæboe, Lars; et al. (2007). "Mobile phone use and risk of glioma in 5 North European countries". *International Journal of Cancer* **120** (8): 1769–75. doi:10.1002/ijc.22503. PMID 17230523.

[33] Lönn, Stefan; Ahlbom, Anders; Hall, Per; Feychting, Maria (2004). "Mobile Phone Use and the Risk of Acoustic Neuroma". *Epidemiology* **15** (6): 653–9. doi:10.1097/01.ede.0000142519.00772.bf. PMID 15475713.

[34] Takebayashi, T; Varsier, N; Kikuchi, Y; Wake, K; Taki, M; Watanabe, S; Akiba, S; Yamaguchi, N (5 February 2008). "Mobile phone use, exposure to radiofrequency electromagnetic field, and brain tumour: a case-control study". *British Journal of Cancer* (London: Nature Publishing Group) **98** (3): 652–659. doi:10.1038/sj.bjc.6604214. PMC 2243154. PMID 18256587. Retrieved 12 March 2008. Lay summary – *Reuters* (5 February 2008). 'Using our newly developed and more accurate techniques, we found no association between mobile phone use and cancer, providing more evidence to suggest they don't cause brain cancer,' Naohito Yamaguchi, who led the research, said.

[35] Hardell, Lennart; Carlberg, Michael; Söderqvist, Fredrik; Mild, Kjell Hansson; Morgan, L. Lloyd (2007). "Long-term use of cellular phones and brain tumours: Increased risk associated with use for ≥10 years". *Occupational and Environmental Medicine* **64** (9): 626–32. doi:10.1136/oem.2006.029751. PMC 2092574. PMID 17409179.

[36] INTERPHONE Study Results update – 7 February 2008

[37] Sage C, Carpenter DO (March 2009). "Public health implications of wireless technologies". *Pathophysiology* **16** (2–3): 233–46. doi:10.1016/j.pathophys.2009.01.011. PMID 19285839.

[38] Hardell L, Carlberg M, Hansson Mild K (March 2009). "Epidemiological evidence for an association between use of wireless phones and tumor diseases". *Pathophysiology* **16** (2–3): 113–22. doi:10.1016/j.pathophys.2009.01.003. PMID 19268551.

[39] Deltour, Isabelle; Johansen, Christoffer; Auvinen, Anssi; Feychting, Maria; Klaeboe, Lars; Schüz, Joachim (16 December 2009). "Time Trends in Brain Tumor Incidence Rates in Denmark, Finland, Norway, and Sweden, 1974–2003". *Journal of the National Cancer Institute* **101** (24): 1721–1724. doi:10.1093/jnci/djp415. PMID 19959779.

[40] Baan, Robert; Yann Grosse; Béatrice Lauby-Secretan; Fatiha El Ghissassi; Véronique Bouvard; Lamia Benbrahim-Tallaa; Neela Guha; Farhad Islami; Laurent Galichet; Kurt Straif (July 2011). "Carcinogenicity of radiofrequency electromagnetic fields". *The Lancet Oncology* **12** (7): 624–626. doi:10.1016/S1470-2045(11)70147-4. PMID 21845765. Retrieved 26 June 2011.

[41] Swerdlow, AJ; Feychting, M; Green, AC; Leeka Kheifets, LK; Savitz, DA; International Commission for Non-Ionizing Radiation Protection Standing Committee on, Epidemiology (November 2011). "Mobile phones, brain tumors, and the interphone study: where are we now?". *Environmental health perspectives* **119** (11): 1534–8. doi:10.1289/ehp.1103693. PMID 22171384.

[42] Repacholi, Michael H.; Lerchl, Alexander; Röösli, Martin; Sienkiewicz, Zenon; Auvinen, Anssi; Breckenkamp, Jürgen; d'Inzeo, Guglielmo; Elliott, Paul; Frei, Patrizia; Heinrich, Sabine; Lagroye, Isabelle; Lahkola, Anna; McCormick, David L.; Thomas, Silke; Vecchia, Paolo (Apr 2012). "Systematic review of wireless phone use and brain cancer and other head tumors". *Bioelectromagnetics* **33** (3): 187–206. doi:10.1002/bem.20716. PMID 22021071.

[43] "U.S. population data show no increase in brain cancer rates during period of expanding cell phone use". Retrieved 12 February 2013.

[44] Coureau, Gaëlle (9 May 2014). "Mobile phone use and brain tumours in the CERENAT case-control study". *Occupational and Environmental Medicine* **71**: 514–522. doi:10.1136/oemed-2013-101754. Retrieved 15 April 2015.

[45] Schönemann, Thomas (6 March 2015). "Höhere Tumorraten durch elektromagnetische Felder". Krebs-Nachrichten. Retrieved 17 March 2015.

[46] Myung, S.-K.; Ju, W.; McDonnell, D. D.; Lee, Y. J.; Kazinets, G.; Cheng, C.-T.; Moskowitz, J. M. (13 October 2009). "Mobile Phone Use and Risk of Tumors: A Meta-Analysis". *Journal of Clinical Oncology* **27** (33): 5565–5572. doi:10.1200/JCO.2008.21.6366.

[47] Luria, Roy; Eliyahu, Ilan; Hareuveny, Ronen; Margaliot, Menachem; Meiran, Nachshon (2009). "Cognitive effects of radiation emitted by cellular phones: The influence of exposure side and time". *Bioelectromagnetics* **30** (3): 198–204. doi:10.1002/bem.20458. PMID 19194860.

[48] Röösli, Martin (June 2008). "Radiofrequency electromagnetic field exposure and non-specific symptoms of ill health: A systematic review". *Environmental Research* **107** (2): 277–287. doi:10.1016/j.envres.2008.02.003. PMID 18359015.

[49] Vijayalaxmi; et al. (2008). "Genetic Damage in Mammalian Somatic Cells Exposed to Radiofrequency Radiation: A Meta-analysis of Data from 63 Publications (1990–2005)". *Radiation Research* **169** (5): 561–574. doi:10.1667/RR0987.1. PMID 18494173.

[50] Ruediger, HW (August 2009). "Genotoxic effects of radiofrequency electromagnetic fields". *Pathophysiology* (Elsevier) **16** (2–3): 67–69. doi:10.1016/j.pathophys.2009.02.002. PMID 19264462.

[51] Harrill, Rob (March 2005). "Wake-up Call". *The University of Washington Alumni Magazine* (March 2005). Retrieved 31 May 2008.

[52] *REFLEX - Risk Evaluation of Potential Environmental Hazards From Low Frequency Electromagnetic Field Exposure Using Sensitive in vitro Methods*. Munich: VERUM Stiftung für Verhalten und Umwelt. 2004. Retrieved 25 May 2011.

[53] Panagopoulos, DJ; Karabarbounis, A; Margaritis, LH (1 December 2004). "Effect of GSM 900 MHz mobile phone radiation on the reproductive capacity of Drosophila melanogaster". *Electromagnetic Biology and Medicine* (London, UK: Taylor & Francis) **23** (1): 29–43. doi:10.1081/JBC-120039350. ISSN 1536-8378. OCLC 87856304. Retrieved 15 January 2008.

[54] Panagopoulos, DJ; Chavdoula, ED; Karabarbounis, A; Margaritis, LH (1 January 2007). "Comparison of bioactivity between GSM 900 MHz and DCS 1800 MHz Mobile Telephony Radiation". *Electromagnetic Biology and Medicine* (London, UK: Informa Healthcare) **26** (1): 33–44. doi:10.1080/15368370701205644. ISSN 1536-8378. OCLC 47815878. PMID 17454081. Retrieved 14 January 2008.

[55] Panagopoulos, DJ; Chavdoula, ED; Nezis, IP; Margaritis, LH (10 January 2007). "Cell death induced by GSM 900 MHz and DCS 1800 MHz mobile telephony radiation". *Mutation Research* (Amsterdam, Netherlands: Elsevier) **626** (1–2): 69–78. doi:10.1016/j.mrgentox.2006.08.008. ISSN 0027-5107. OCLC 109920000. PMID 17045516. Retrieved 15 January 2008. Our present results suggest that the decrease in oviposition previously reported, is due to degeneration of large numbers of egg chambers after DNA fragmentation of their constituent cells, induced by both types of mobile telephony radiation. Induced cell death is recorded for the first time, in all types of cells constituting an egg chamber...

[56] De Iuliis, Geoffry N.; Rhiannon J. Newey; Bruce V. King; R. John Aitken (31 July 2009). Zhang, Baohong, ed. "Mobile Phone Radiation Induces Reactive Oxygen Species Production and DNA Damage in Human Spermatozoa *In Vitro*".

PLoS ONE (Callaghan, New South Wales, Australia) **4(7)** (e6446): e6446. doi:10.1371/journal.pone.0006446. ISSN 1932-6203. PMC 2714176. PMID 19649291.

[57] Haarala, C; Takio F; Rintee T; Laine M; Koivisto M; Revonsuo A; Hämäläinen H (May 2007). "Pulsed and continuous wave mobile phone exposure over left versus right hemisphere: effects on human cognitive function". *Bioelectromagnetics* (Wiley-Liss, Inc) **28** (4): 289–95. doi:10.1002/bem.20287. PMID 17203481.

[58] Borbély, AA; Huber R; Graf T; Fuchs B; Gallmann E; Achermann P (19 November 1999). "Pulsed high-frequency electromagnetic field affects human sleep and sleep electroencephalogram". *Neuroscience Letters* (East Park, Ireland: Elsevier Science Ireland) **275** (3): 207–10. doi:10.1016/S0304-3940(99)00770-3. PMID 10580711.

[59] Huber, R; Graf T; Cote KA; Wittmann L; Gallmann E; Matter D; Schuderer J; Kuster N; Borbély AA; Achermann P (20 October 2000). "Exposure to pulsed high-frequency electromagnetic field during waking affects human sleep EEG". *NeuroReport* (Lippincott Williams & Wilkins, Inc) **11** (15): 3321–5. doi:10.1097/00001756-200010200-00012. PMID 11059895.

[60] Huber, R; Treyer V; Borbély AA; Schuderer J; Gottselig JM; Landolt HP; Werth E; Berthold T; Kuster N; Buck A; Achermann P (December 2002). "Electromagnetic fields, such as those from mobile phones, alter regional cerebral blood flow and sleep and waking EEG". *Journal of sleep research* (Wiley-Liss, Inc) **11** (4): 289–95. doi:10.1046/j.1365-2869.2002.00314.x. PMID 12464096.

[61] Huber, R; Treyer V; Schuderer J; Berthold T; Buck A; Kuster N; Landolt HP; Achermann P (February 2005). "Exposure to pulse-modulated radio frequency electromagnetic fields affects regional cerebral blood flow". *The European Journal of Neuroscience* (Wiley-Liss, Inc) **21** (4): 1000–6. doi:10.1111/j.1460-9568.2005.03929.x. PMID 15787706.

[62] Hung, CS; Anderson C; Horne, JA; McEvoy, P (21 June 2007). "Mobile phone 'talk-mode' signal delays EEG-determined sleep onset". *Neuroscience Letters* (East Park, Ireland: Elsevier Science Ireland) **421** (1): 82–6. doi:10.1016/j.neulet.2007.05.027. ISSN 0304-3940. PMID 17548154.

[63] Andrzejak, R; Poreba R; Poreba M; Derkacz A; Skalik R; Gac P; Beck B; Steinmetz-Beck A; Pilecki W (August 2008). "The influence of the call with a mobile phone on heart rate variability parameters in healthy volunteers". *Industrial health* (National Institute of Industrial Health) **46** (4): 409–17. doi:10.2486/indhealth.46.409. PMID 18716391.

[64] Krause, CM; Pesonen M; Haarala Björnberg C; Hämäläinen H (May 2007). "Effects of pulsed and continuous wave 902 MHz mobile phone exposure on brain oscillatory activity during cognitive processing". *Bioelectromagnetics* (Wiley-Liss, Inc) **28** (4): 296–308. doi:10.1002/bem.20300. PMID 17203478.

[65] Papageorgiou, CC; Nanou ED; Tsiafakis VG; Kapareliotis E; Kontoangelos KA; Capsalis CN; Rabavilas AD; Soldatos CR (10 April 2006). "Acute mobile phone effects on pre-attentive operation". *Neuroscience Letters* (East Park, Ireland: Elsevier Science Ireland) **397** (1–2): 99–103. doi:10.1016/j.neulet.2005.12.001. PMID 16406308.

[66] Kramarenko, AV; Tan U (July 2003). "Effects of high-frequency electromagnetic fields on human EEG: a brain mapping study". *The International journal of neuroscience* (Taylor and Francis) **113** (7): 1007–19. doi:10.1080/00207450390220330. PMID 12881192.

[67] D'Costa, H; Trueman G; Tang L; Abdel-rahman U; Abdel-rahman W; Ong K; Cosic I (December 2003). "Human brain wave activity during exposure to radiofrequency field emissions from mobile phones". *Australas Phys Eng Sci Med* (Australasian College Of Physical Scientists In Medicine) **26** (4): 162–7. doi:10.1007/BF03179176. ISSN 0158-9938. PMID 14995060.

[68] Krause, CM; Björnberg CH; Pesonen M; Hulten A; Liesivuori T; Koivisto M; Revonsuo A; Laine M; Hämäläinen H (June 2006). "Mobile phone effects on children's event-related oscillatory EEG during an auditory memory task". *International journal of radiation biology* (Taylor and Francis) **82** (6): 443–50. doi:10.1080/09553000600840922. PMID 16846979.

[69] Aalto, S; Haarala C; Brück A; Sipilä H; Hämäläinen H; Rinne JO (July 2006). "Mobile phone affects cerebral blood flow in humans". *J Cereb Blood Flow Metab* (Nature Publishing Group) **26** (7): 885–90. doi:10.1038/sj.jcbfm.9600279. PMID 16495939.

[70] Bachmann, M; Lass J; Kalda J; Säkki M; Tomson R; Tuulik V; Hinrikus H; Kalda, J; Säkki, M; Tomson, R; Tuulik, V; Hinrikus, H (2006). "Integration of differences in EEG analysis reveals changes in human EEG caused by microwave". *Conf Proc IEEE Eng Med Biol Soc* (IEEE Service Center) **1**: 1597–600. doi:10.1109/IEMBS.2006.259234. PMID 17946053.

[71] "Cell Phone Use in Pregnancy May Cause Behavioral Disorders in Offspring, Mouse Study Suggests". Science Daily. Retrieved 1 April 2012.

[72] Fejes I, Zavaczki Z, Szollosi J, Koloszar S, Daru J, Kovacs L, et al. (2005). "Is there a relationship between cell phone use and semen quality?". *Arch Androl* **51** (5): 385–93.

[73] Kilgallon SJ, Simmons LW (2005). "Image content influences men's semen quality". *Biol Lett* **1** (3): 253–5. doi:10.1098/rsbl.2005.0324.

[74] Wdowiak A, Wdowiak L, Wiktor H (2007). "Evaluation of the effect of using mobile phones on male fertility". *Ann Agric Environ Med* **14** (1): 169–72.

[75] Agarwal A, Deepinder F, Sharma RK, Ranga G, Li J (2008). "Effect of cell phone usage on semen analysis in men attending infertility clinic: an observational study". *Fertil Steril* **89** (1): 124–8. doi:10.1016/j.fertnstert.2007.01.166.

[76] Gutschi T, Mohamad Al-Ali B, Shamloul R, Pummer K, Trummer H (2011). "Impact of cell phone use on men's semen parameters". *Andrologia* **43** (5): 312–6. doi:10.1111/j.1439-0272.2011.01075.x.

[77] Davoudi M, Brossner C, Kuber W. 2002. The influence of electromagnetic waves on sperm motility. Journal für Urologie und Urogynäkologie 19: 19-22.

[78] Erogul O, Oztas E, Yildirim I, Kir T, Aydur E, Komesli G, et al. (2006). "Effects of electromagnetic radiation from a cellular phone on human sperm motility: an in vitro study". *Arch Med Res* **37** (7): 840–3. doi:10.1016/j.arcmed.2006.05.003.

[79] Agarwal A, Desai NR, Makker K, Varghese A, Mouradi R, Sabanegh E, et al. (2009). "Effects of radiofrequency electromagnetic waves (RF-EMW) from cellular phones on human ejaculated semen: an in vitro pilot study". *Fertil Steril* **92** (4): 1318–25. doi:10.1016/j.fertnstert.2008.08.022. PMID 18804757.

[80] De Iuliis GN, Newey RJ, King BV, Aitken RJ (2009). "Mobile phone radiation induces reactive oxygen species production and DNA damage in human spermatozoa in vitro". *PLOS ONE* **4** (7): e6446. doi:10.1371/journal.pone.0006446. PMC 2714176. PMID 19649291.

[81] Falzone N, Huyser C, Becker P, Leszczynski D, Franken DR (2011). "The effect of pulsed 900-MHz GSM mobile phone radiation on the acrosome reaction, head morphometry and zona binding of human spermatozoa". *Int J Androl* **34** (1): 20–6.

[82] http://www.ewg.org/cell-phone-radiation-damages-sperm-studies-find

[83] Santini, R; Santini, P; Danze, JM; LeRuz, P; Seigne, M (January 2003). "Survey Study of People Living in the Vicinity of Cellular Phone Base Stations". *Electromagnetic Biology and Medicine* (London: Informa Healthcare) **22** (1): 41–49. doi:10.1081/JBC-120020353. OCLC 88891277. Retrieved 9 February 2008.

[84] Navarro, Enrique A; Segura, J; Portolés, M; Gómez-Perretta de Mateo, Claudio (December 2003). "The Microwave Syndrome: A Preliminary Study in Spain". *Electromagnetic Biology and Medicine* (London: Informa Healthcare) **22** (2): 161–169. doi:10.1081/JBC-120024625. OCLC 89106315. Retrieved 9 February 2008.
Oberfeld, Gerd; Navarro, Enrique A; Portoles, Manuel; Maestu, Ceferino; Gomez-Perretta, Claudio (2004). "The Microwave Syndrome: Further Aspects of a Spanish Study". In Kostarakis, P. *Biological effects of EMFs : Proceedings, Kos, Greece, 4–8 October 2004, 3rd International Workshop*. Ioannina, Greece: Electronics, Telecom & Applications Laboratory, Physics Dept., University of Ioannina : Institute of Informatics & Telecommunications, N.C.S.R. "Demokritos". ISBN 960-233-152-6.

[85] Abdel-Rassoul, G; Abou El-Fateh, O; Abou Salem, M; Michael, A; Farahat, F; El-Batanouny, M; Salem, E

(March 2007). "Neurobehavioral effects among inhabitants around mobile phone base stations" (PDF). *NeuroToxicology* (New York, NY: Elsevier Science) **28** (2): 434–40. doi:10.1016/j.neuro.2006.07.012. OCLC 138574974. PMID 16962663. Retrieved 10 February 2008.

[86] Bortkiewicz, A; Zmyślony, M; Szyjkowska, A; Gadzicka, E (2004). "Subjective symptoms reported by people living in the vicinity of cellular phone base stations: review". *Medycyna pracy* (in Polish) (Warsaw: Państwowy Zakład Wydawnictw Lekarskich) **55** (4): 345–352. ISSN 0465-5893. OCLC 108011911. PMID 15620045. BL Shelfmark: 5536.020000.

[87] Hutter, H-P; H Moshammer; P Wallner; M Kundi (1 May 2006). "Subjective symptoms, sleeping problems, and cognitive performance in subjects living near mobile phone base stations". *Occupational and Environmental Medicine* (London, UK: the BMJ Publishing Group) **63** (5): 307–313. doi:10.1136/oem.2005.020784. OCLC 41236398. PMC 2092490. PMID 16621850. Retrieved 7 January 2008.

[88] Neubauer; et al. (2007). "Feasibility of future epidemiological studies on possible health effects of mobile phone base stations". *Bioelectromagnetics* **28** (3): 224–230. doi:10.1002/bem.20298. PMID 17080459.

[89] Regel SJ, Negovetic S, Röösli M, et al. (2006). "UMTS Base Station-like Exposure, Well-Being, and Cognitive Performance". *Environmental Health Perspectives* **114** (8): 8. doi:10.1289/ehp.8934. PMC 1552030. PMID 16882538.

[90] Eltiti, S; Wallace, D; Ridgewell, A; Zougkou, K; Russo, R; Sepulveda, F; Mirshekar-Syahkal, D; Rasor, P; Deeble, R; Fox, E (November 2007). "Does short-term exposure to mobile phone base station signals increase symptoms in individuals who report sensitivity to electromagnetic fields? A double-blind randomized provocation study". *Environ Health Perspect* **115** (11): 1603–1608. doi:10.1289/ehp.10286. OCLC 183843559. PMC 2072835. PMID 18007992. Lay summary – *Study finds health symptoms aren't linked to mast emissions University of Essex* (25 July 2007).

[91] http://www.afsset.fr/index.php?pageid=712&parentid=424 page 37

[92] Téléphonie mobile et santé, Rapport à l'Agence Française de Sécurité Sanitaire Environnementale, 21 March 2003 at http://www.afsset.fr/index.php?pageid=712&parentid=424

[93] Téléphonie mobile et santé, Rapport du groupe d'experts, l'Agence Française de Sécurité Sanitaire Environnementale, April 2005 at http://www.afsset.fr/index.php?pageid=712&parentid=424

[94] "Radiofréquences : actualisation de l'expertise (2009)", l'Agence Française de Sécurité Sanitaire Environnementale, April 2005 at http://www.afsset.fr/index.php?pageid=712&parentid=424

[95] Moulder, JE; Erdreich, LS; Malyapa, RS; Merritt, J; Pickard, WF; Vijayalaxmi (May 1999). "Cell phones and cancer: what is the evidence for a connection?". *Radiation Research* (New York: Academic Press) **151** (5): 513–531. doi:10.2307/3580028. ISSN 0033-7587. JSTOR 3580028. OCLC 119963820. PMID 10319725. Retrieved 10 February 2008.

[96] "International Commission for Non-Ionizing Radiation Protection home page". Retrieved 7 January 2008.

[97] "Anforderungen nach NISV: Mobilfunkanlagen" [Specifications of the Regulation on Non-Ionizing Radiation: Mobile Telephone Installations] (in German). Bundesamt für Umwelt [Swiss Federal Environment Ministry]. 13 March 2009. Retrieved 20 January 2010.

[98] Wright v. Motorola, Inc. et al., No95-L-04929

[99] *Christopher Newman, et al. v Motorola, Inc., et al.* (United States District Court for the District of Maryland) ("Because no sufficiently reliable and relevant scientific evidence in support of either general or specific causation has been proffered by the plaintiffs, as explained below, the defendants' motion will be granted and the plaintiffs' motion will be denied."). Text

[100] Barstad, Stine (18 February 2009). "Kunne ikke bevise at strålingen var ufarlig". *Aftenposten* (in Norwegian). Archived from the original on 25 May 2009. Retrieved 25 May 2009.

[101] *Residents living next to a phone mast vs. the mobile phone company Bouygues Telecom* (Versailles Court of Appeal 4 February 2009). Text

[102] "Cassazione Civile, 12 ottobre 2012, n. 17438 – Uso di telefoni nel corso dell'attività lavorativa e patologia tumorale". *www.leggioggi.it* (in Italian). Retrieved 16 March 2015.

[103] "Italian Supreme Court Rules Cell Phones Can Cause Cancer" (Press release). Center for Family and Community Health. 19 October 2012. Retrieved 16 March 2015.

[104] "Italy court ruling links mobile phone use to tumour". *Reuters*. 19 October 2012.

[105] "Bereaved dad seeks curbs on mobile towers - The Times of India". *The Times Of India*.

[106] http://www.iarc.fr/en/media-centre/pr/2011/pdfs/pr208_E.pdf

[107] http://www.ee.iitb.ac.in/~{}mwave/Cell-tower-rad-report-WB-Environ-Oct2011.pdf

[108] "Notice to central, Delhi governments on removal of mobile towers". *Daily News* (New York). Archived from the original on 2013-06-26.

[109] "Electromagnetic Fields and Public Health - Cautionary Policies". *World Health Organization Backgrounder*. World Health Organization. March 2000. Retrieved 1 February 2008.

[110] Wiedemann; et al. (2006). "The Impacts of Precautionary Measures and the Disclosure of Scientific Uncertainty on EMF Risk Perception and Trust". *Journal of Risk Research* **9** (4): 361–372. doi:10.1080/13669870600802111.

[111] Poumadère M., Perrin A. (2013). "Risk Assessment of Radiofrequencies and Public Information". *Journal of Risk Analysis and Crisis Response* **3** (1): 3–12. doi:10.2991/jrarc.2013.3.1.1.

[112] "Téléphones mobiles : santé et sécurité" (in French). Le ministère de la santé, de la jeunesse et des sports. 2 January 2008. Retrieved 19 January 2008. Lay article in (English) making comment at Gitlin, Jonathan M. (3 January 2008). "France: Beware excessive cell phone use—despite lack of data". Ars Technica. Retrieved 19 January 2008.

[113] "Precaution regarding electromagnetic fields". Federal Office for Radiation Protection. 7 December 2007. Retrieved 19 January 2008.

[114] "Exponering" (in Swedish). Swedish Radiation Protection Authority. February 2006. Retrieved 19 January 2008.

[115] "UK consumer group: Hands-free phone kits boost radiation exposure". *cnn.com* (Cable News Network). 2 November 2000. Retrieved 19 January 2008.

[116] Manning, MI and Gabriel, CHB, SAR tests on mobile phones used with and without personal hands-free kits, SARtest Report 0083 for the DTI, July 2000 (PDF) at http://straff-x.com/SAR-Hands-Free-Kits-July-2000.pdf

[117] Téléphonie mobile & santé, Report for l"Agence française de sécurité sanitaire environnementale (Afsse), June 2005 at http://www.afsse.fr/index.php?pageid=671&parentid=619#

[118] "Bead 'slashes mobile radiation'". BBC News. 25 January 2005. Retrieved 17 March 2009.

[119] For example, Finland "Radiation and Nuclear Safety Authority: Children's mobile phone use should be limited". Finnish Radiation and Nuclear Safety Authority (STUK). 7 January 2009. Retrieved 20 January 2010. and France "Téléphone mobile, DAS et santé" [Mobile telephones, SAR and health] (PDF). *Votre enfant et le téléphone mobile [Your child and mobile telephony].* Association Française des Opérateurs Mobiles (AFOM)[French Mobile Phone Operators' Association] et l'Union Nationale des Associations Familiales (UNAF) [National Federation of Family Associations]. 31 January 2007. Retrieved 20 January 2010.

[120] Gandhi, Om P.; Morgan, L. Lloyd; de Salles, Alvaro Augusto; Han, Yueh-Ying; Herberman, Ronald B.; Davis, Devra Lee (October 14, 2011). "Exposure Limits: The underestimation of absorbed cell phone radiation, especially in children". *Electromagnetic Biology and Medicine* **31** (1): 34–51. doi:10.3109/15368378.2011.622827. ISSN 1536-8378. Retrieved 2015-04-25.

1.2.10 External links

- Summary and full text of "Possible effects of Electromagnetic Fields (EMF) on Human Health", the 2007 scientific assessment of the European Commission's SCENIHR (Scientific Committee on Emerging and Newly Identified Health Risks).

- WHO International EMF Program

- Independent Expert Group on Mobile Phones (IEGMP), UK

- FDA Cell Phone Facts

- FCC Radio Frequency Safety

- Medline Plus, by US National Library of Medicine and National Institutes of Health (NIH)

- GSM Association: Health

- Public health and electromagnetic fields: Overview of European Commission activities

Chapter 2

Mobile Phone Radiation Info

2.1 Bioinitiative Report

The BioInitiative Report is a report on the relationship between the electromagnetic fields (EMF) associated with powerlines and wireless devices and health. It was self-published online, without peer review, on 31 August 2007, by a group "of 14 scientists, researchers, and public health policy professionals". The BioInitiative Report states that it is an examination of the controversial health risks of electromagnetic fields and radiofrequency radiation.[1] Some updated BioInitiative material was published in a journal in an issue guest-edited by one of the members of the group,[2] and a 2012 version of the report was released on 7 January 2013.[3] It has been heavily criticized by independent and governmental research groups for its lack of balance.

2.1.1 History

In 2006, at the Bioelectromagnetics Society's annual meeting, there was a mini-symposium on electromagnetic fields and radiofrequency radiation to present the science showing biological effects, and the precautionary measures taken by countries around the world. The Bioinitiative Working Group grew out of this conference and decided to write a report on the science and health risks to alert people who could translate the science into public policy. From October 2006 to August 2007, 14 scientists and public health experts worked to come up with recommendations for the Bioinitiative Report.[4]

Since 2007, some of the material was revised, updated and submitted for peer-reviewed publication and published in the August 2009 issue of *Pathophysiology*, an issue guest-edited by Martin Blank, one of the three members of the BioInitiative Organizing Committee.[2]

An updated 2012 version of the report was released on 7 January 2013.[3]

2.1.2 Criticism

The following government health authorities and independent expert groups have reviewed the BioInitiative Report and made the following comments on the merit of its claims.

Health Council of the Netherlands

The Health Council of the Netherlands reviewed the BioInitiative report in September 2008 and concluded it is a selective review of existing research and does not present a balanced analysis considering the relative scientific quality of different studies. Some of the shortcomings identified included that the report made claims which lacked scientific basis and false claims.

In 2008 they concluded:

> In view of the way the BioInitiative report was compiled, the selective use of scientific data and the other shortcomings mentioned above, the Committee concludes that the BioInitiative report is not an objective and balanced reflection of the current state of scientific knowledge.[5]

Australian Centre for Radiofrequency Bioeffects Research (ACRBR)

In December 2008 the Australian Centre for Radiofrequency Bioeffects Research (ACRBR) reviewed the BioInitiative Report and concluded:

> Overall we think that the BioInitiative Report does not progress science, and would agree with the Health Council of the Netherlands that the BioInitiative Report is "not an objective and balanced reflection of the current state of scientific knowledge". As it stands it merely provides a set of views that are not consistent with the consensus of science, and it does not provide an analysis

that is rigorous-enough to raise doubts about the scientific consensus.

The ACRBR also points out there are statements in the report that do not accord with the standard view of science, and the report does not provide a reasonable account of why we should reject the standard view in favour of the views espoused in the report.

The ACRBR also noted that the state of science in this area is continually being debated and updated by a number of expert bodies composed of the leading experts in this field and strongly urged people to consult these views for a balanced assessment of the research.[6]

European Commission's EMF-NET

The European Commission's EMF-NET coordination group for investigating the impact of electromagnetic fields on health made the following comments in October 2007 regarding the BioInitiative Report:

> There is a lack of balance in the report; no mention is made in fact of reports that do not concur with authors' statements and conclusions. The results and conclusions are very different from those of recent national and international reviews on this topic... If this report were to be believed, EMF would be the cause of a variety of diseases and subjective effects...[7]

Institute of Electrical and Electronics Engineers (IEEE) Committee on Man and Radiation (COMAR)

The Institute of Electrical and Electronics Engineers (IEEE) Committee on Man and Radiation (COMAR) reviewed the BioInitiative Report in 2009. They concluded:

> ...that the weight of scientific evidence in the RF bioeffects literature does not support the safety limits recommended by the BioInitiative group. For this reason, COMAR recommends that public health officials continue to base their policies on RF safety limits recommended by established and sanctioned international organizations such as the Institute of Electrical and Electronics Engineers International Committee on Electromagnetic Safety and the International Commission on Non-Ionizing Radiation Protection, which is formally related to the World Health Organization.[8]

German Federal Office for Radiation Protection

The German Federal Office for Radiation Protection (BfS) commented in October 2007 on a newsmagazine TV show on the German network ARD that featured the BioInitiative Report shortly after its release. They said:

> The BfS conducted a preliminary review of the so-called "BioInitiative Report" immediately after its release and concluded that it had clear scientific shortcomings. In particular, it has undertaken to combine the health effects of low- and high-frequency fields that are not technically possible. The overwhelming majority of studies underpinning the report are not new: they already have been taken into account in the determination of currently applicable standards.[9]

French Agency for Environmental and Occupational Health Safety

The French Agency for Environmental and Occupational Health Safety (*Agence française de sécurité sanitaire de l'environnement et du travail*, AFSSET) analysed the contents of the BioInitiative Report and in October 2009 said:

> ...the different chapters of the report are of uneven editing style and quality. Some sections do not present scientific data in a balanced fashion, do not analyze the quality of the articles cited, or reflect the personal opinions of their authors ..., [the report] is tinged with conflicts of interest in several chapters, does not reflect a collective effort, and is written in militant style.[10]

Indian Council of Medical Research

The Indian Council of Medical Research reviewed the 2012 version of Bioinitiative Report in February 2013 and said:

> ...on critical examination of the Bioinitiative 2012 Report, has observed that the report is not based on multi disciplinary weight – of evidence method leads to a scientifically sound judgment & objective and there is no balanced reflection of the current state of scientific knowledge. However, the evidence given in the report cannot be ignored and hence, need further investigation in this area.[11][12]

Other

In the March/April 2008 newsletter of the Bioelectromagnetics Society, publishers of the journal *Bioelectromagnetics* and to which several BioInitiative Report contributors belong, a commentary noted "...analysis by good theoretical physicists suggests that nothing is going to happen but the deposition of additional energy that, if sufficient, can elevate tissue temperature. But physicists don't know everything so we turn to the biologists and find that an analysis of the biological database reveals no consistently reproducible (independent) LLNT effect after about 50 or 60 years of research."[13]

2.1.3 See also

- COSMOS cohort study

- Electromagnetic hypersensitivity

- Electromagnetic radiation and health

- Mobile phone radiation and health

- Wireless electronic devices and health

2.1.4 References

[1] Bioinitiative Report

[2] Blank, Martin (August 2009). "Preface". *Pathophysiology* **16** (2–3): 67–69. doi:10.1016/j.pathophys.2009.02.002. PMID 19264462. And "List of BioInitiative Participants". BioInitiative Report. Archived from the original on 22 April 2008. Retrieved 2010-01-20.

[3] "BioInitiative 2012 Report Issues New Warnings on Wireless and EMF" (PDF). Bionitiative Report. 7 January 2013. Retrieved 2013-01-20.

[4] "The BioInitiative Report - Biological Standards for Wireless". Retrieved 2008-06-07.

[5] "BioInitiative report (publication no. 2008/17E)". Health Council of the Netherlands. 1 September 2008. Retrieved 2010-01-20.

[6] Rodney, Croft; Abramson, Michael; Cosic, Irena; Finnie, John; McKenzie, Ray; Wood, Andrew (18 December 2008). "ACRBR Position Statement on *BioInitiative Report*" (PDF). Australian Centre for Radiofrequency Bioeffects Research. Retrieved 2010-01-20.

[7] "Comments on the BioInitiative Working Group Report (BioInitiative Report)" (PDF). EMF-NET. 30 October 2007. Retrieved 2010-01-20.

[8] Committee on Man and Radiation (COMAR) (October 2009). "COMAR technical information statement: expert reviews on potential health effects of radiofrequency electromagnetic fields and comments on the bioinitiative report". *Health Physics* **97** (4): 348–356. doi:10.1097/HP.0b013e3181adcb94. PMID 19741364.

[9] "Stellungnahme zur Sendung "Bei Anruf Hirntumor?" von Report Mainz vom 29.10.2007" [Position on the Broadcast by Report Mainz "Brain Tumours by Telephone?" of 29 October 2007] (in German). Bundesamt für Strahlenschutz [Federal Office for Radiation Protection]. 30 October 2007. Archived from the original on 2013-05-28. Retrieved 2010-01-20. Das BfS hat den sogenannten „BioInitiative Report" unmittelbar nach dessen Publikation einer ersten Prüfung unterzogen und festgestellt, dass er klare wissenschaftliche Schwächen aufweist: Insbesondere werden Vermischungen der gesundheitlichen Wirkungen von niederfrequenten und hochfrequenten Feldern vorgenommen, die fachlich nicht zulässig sind. Die überwiegende Mehrzahl der dem Report zugrunde liegenden Studien ist nicht neu: Sie wurden bei der Festlegung der derzeit gültigen Grenzwerte bereits berücksichtigt.

[10] "5.3.1 BioInitiative" (PDF). *Mise à jour de l'expertise relative aux radiofréquences - Saisine n°2007/007 [Update on the state of radiofrequency research - Reference #2007/007]* (in French). Agence française de sécurité sanitaire de l'environnement et du travail [French Agency for Environmental and Occupational Health Safety]. October 2009. pp. 322–326. ...les différents chapitres du rapport sont de rédaction et de qualité inégales. Certains articles ne présentent pas les données scientifiques disponibles de manière équilibrée, n'analysent pas la qualité des articles cités ou reflètent les opinions ou convictions personnelles de leurs auteurs (...), il revêt des conflits d'intérêts dans plusieurs chapitres, ne correspond pas à une expertise collective et est écrit sur un registre militant.

[11] "Study on Radiation From Mobile Towers and Cell Phones". Indian Council of Medical Research. 22 February 2013. Retrieved 2013-03-08.

[12] "Study on Radiation From Mobile Towers and Cell Phones" (Press release). Press Information Bureau. 2013-02-22. 92395. Retrieved 2014-12-09.

[13] Swicord, Mays (March–April 2008). "Retirement and RF Biological Effects" (PDF). *Bioelectromagnetics Newsletter* (201): 7. Retrieved 2011-03-14.

2.1.5 External links

- BioInitiative Report

- World Health Organization EMF Project

- International Commission on Non Ionizing Radiation

- European Environment Agency

2.2 Dielectric heating

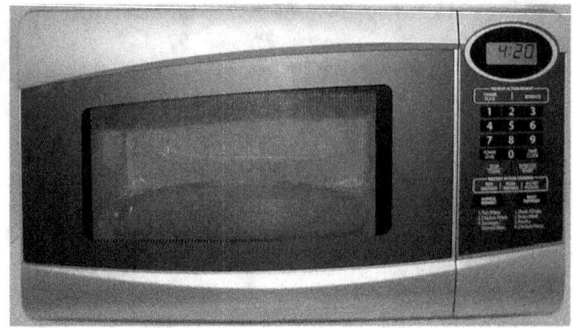

A microwave oven, which uses dielectric heating to cook food

Dielectric heating, also known as **electronic heating**, **RF heating**, **high-frequency heating** is the process in which a high-frequency alternating electric field, or radio wave or microwave electromagnetic radiation heats a dielectric material. At higher frequencies, this heating is caused by molecular dipole rotation within the dielectric.

RF dielectric heating at intermediate frequencies, due to its greater penetration over microwave heating, shows greater promise than microwave systems as a method of very rapidly heating and uniformly preparing certain food items, and also killing parasites and pests in certain harvested crops.[1]

2.2.1 Mechanism

Molecular rotation occurs in materials containing polar molecules having an electrical dipole moment, with the consequence that they will align themselves in an electromagnetic field. If the field is oscillating, as it is in an electromagnetic wave or in a rapidly oscillating electric field, these molecules rotate continuously aligning with it. This is called dipole rotation, or dipolar polarisation. As the field alternates, the molecules reverse direction. Rotating molecules push, pull, and collide with other molecules (through electrical forces), distributing the energy to adjacent molecules and atoms in the material. Once distributed, this energy appears as heat.

Temperature is related to the average kinetic energy (energy of motion) of the atoms or molecules in a material, so agitating the molecules in this way increases the temperature of the material. Thus, dipole rotation is a mechanism by which energy in the form of electromagnetic radiation can raise the temperature of an object. There are also many other mechanisms by which this conversion occurs.

Dipole rotation is the mechanism normally referred to as dielectric heating, and is most widely observable in the microwave oven where it operates most efficiently on liquid water, and much less so on fats and sugars. This is because fats and sugar molecules are far less polar than water molecules, and thus less affected by the forces generated by the alternating electromagnetic fields. Outside of cooking, the effect can be used generally to heat solids, liquids, or gases, provided they contain some electric dipoles.

Dielectric heating involves the heating of electrically insulating materials by dielectric loss. A changing electric field across the material causes energy to be dissipated as the molecules attempt to line up with the continuously changing electric field. This changing electric field may be caused by an electromagnetic wave propagating in free space (as in a microwave oven), or it may be caused by a rapidly alternating electric field inside a capacitor. In the latter case there is no freely propagating electromagnetic wave, and the changing electric field may be seen as analogous to the electric component of an antenna near field. In this case, although the heating is accomplished by changing the electric field inside the capacitive cavity at radio-frequency (RF) frequencies, no actual radio waves are either generated or absorbed. In this sense, the effect is the direct electrical analog of magnetic induction heating, which is also near-field effect (and also does not involve classical radio waves).

Frequencies in the range of 10–100 MHz are necessary to cause efficient dielectric heating, although higher frequencies work equally well or better, and in some materials (especially liquids) lower frequencies also have significant heating effects, often due to more unusual mechanisms. For example, in conductive liquids such as salt water, "ion-drag" causes heating, as charged ions are "dragged" more slowly back and forth in the liquid under influence of the electric field, striking liquid molecules in the process and transferring kinetic energy to them, which is eventually translated into molecular vibrations and thus into thermal energy.

Dielectric heating at low frequencies, as a near-field effect, requires a distance from electromagnetic radiator to absorber of less than about 1/6th of a wavelength ($\lambda/2\pi$) of the source frequency. It is thus a contact process or near-contact process, since it usually sandwiches the material to be heated (usually a non-metal) between metal plates that set up to form what is effectively a very large capacitor, with the material to be heated acting as the dielectric inside the capacitor. However, actual electrical contact is not necessary for heating a dielectric inside a capacitor, as the electric fields that form inside a capacitor subjected to a voltage do not require electrical contact of the capacitor plates with the dielectric (non-conducting) material between the plates. Because lower frequency electrical fields penetrate nonconductive materials far more deeply than do microwaves, heating pockets of water and organisms deep inside dry materials like wood, it can be used to rapidly heat and prepare many non-electrically conducting food and agricultural items, so long as they fit between the capacitor plates.

At very high frequencies, the wavelength of the electromagnetic field becomes shorter than the distance between the metal walls of the heating cavity, or than the dimensions of the walls themselves. This is the case inside a microwave oven. In such cases, conventional far-field electromagnetic waves form (the cavity no longer acts as a pure capacitor, but rather as an antenna), and are absorbed to cause heating, but the dipole-rotation mechanism of heat deposition remains the same. However, microwaves are not efficient at causing the heating effects of low frequency fields that depend on slower molecular motion, such as those caused by ion-drag.

2.2.2 Power

Dielectric heating must be distinguished from Joule heating of conductive media, which is caused by induced electric currents in the media. For dielectric heating, the generated power density per volume is given by:

$$Q = \omega \cdot \varepsilon_r'' \cdot \varepsilon_0 \cdot E^2,$$

where ω is the angular frequency of the exciting radiation, ε_r'' is the imaginary part of the complex relative permittivity of the absorbing material, ε_0 is the permittivity of free space and E the electric field strength. The imaginary part of the (frequency-dependent) relative permittivity is a measure for the ability of a dielectric material to convert electromagnetic field energy into heat.

If the conductivity σ of the material is small, or the frequency is high, such that $\sigma \ll \omega\varepsilon$ (with $\varepsilon = \varepsilon_r'' \cdot \varepsilon_0$), then dielectric heating is the dominant mechanism of loss of energy from the electromagnetic field into the medium.

2.2.3 Penetration

Microwave frequencies penetrate conductive materials, including semi-solid substances like meat and living tissue, to a distance defined by the skin effect. The penetration essentially stops where all the penetrating microwave energy has been converted to heat in the tissue. Microwave ovens used to heat food are not set to the frequency for optimal absorption by water. If that was so, then the piece of food or liquid in question would absorb all microwave radiation in its outer layer, leading to a cool, unheated centre and a superheated surface. Instead, the frequency selected allows energy to penetrate deeper into the heated food. The frequency of a household microwave oven is 2.45 GHz, while the frequency for optimal absorbency by water is around 10 GHz. [2]

2.2.4 Use of RF electric fields in dielectric heating

The use of high-frequency electric fields for heating dielectric materials had been proposed in the 1930s. For example, U.S. Patent 2,147,689 (application by Bell Telephone Laboratories, dated 1937) states "*This invention relates to heating systems for dielectric materials and the object of the invention is to heat such materials uniformly and substantially simultaneously throughout their mass. It has been proposed therefore to heat such materials simultaneously throughout their mass by means of the dielectric loss produced in them when they are subjected to a high voltage, high frequency field.*" This patent proposed radio frequency (RF) heating at 10 to 20 megahertz (wavelength 15 to 30 meters).[3] Such wavelengths were far longer than the cavity used, and thus made use of near-field effects and not electromagnetic waves. (Commercial microwave ovens use wavelengths only 1% as long.)

In agriculture, RF dielectric heating has been widely tested and is increasingly used as a way to kill pests in certain food crops after harvest, such as walnuts still in the shell. Because RF heating can heat foods more uniformly than is the case with microwave heating, RF heating holds promise as a way to process foods quickly.[1]

In medicine, the RF heating of body tissues is a kind of diathermia or diathermy.[4]

2.2.5 Microwave heating

Microwave heating, as distinct from RF heating, is a subcategory of dielectric heating at frequencies above 100 MHz, where an electromagnetic wave can be launched from a small dimension emitter and guided through space to the target. Modern microwave ovens make use of electromagnetic waves (microwaves) with electric fields of much higher frequency and shorter wavelength than RF heaters. Typical domestic microwave ovens operate at 2.45 GHz, but 915 MHz ovens also exist. This means that the wavelengths employed in microwave heating are 12 or 33 cm (4.7 or 13 inch). This provides for highly efficient, but less penetrative, dielectric heating.

Although a capacitor-like set of plates can be used at microwave frequencies, they are not necessary, since the microwaves are already present as far field type EM radiation, and their absorption does not require the same proximity to a small antenna as does RF heating. The material to be heated (a non-metal) can therefore simply be placed in the path of the waves, and heating takes place in a non-contact process which does not require capacitative conductive plates.

2.2.6 Microwave Volumetric Heating

Microwave Volumetric Heating is a commercially available method of heating liquids, suspensions, or solids in a continuous flow on an industrial scale. Microwave Volumetric Heating has a greater penetration depth, of up to 42 mm, which is an even penetration through the entire volume of the flowing product. This is advantageous in commercial applications where increased shelf-life can be achieved, with increased microbial kill at temperatures 10-15 °C lower than when using conventional heating systems.

Application for Microwave Volumetic Heating:

- Pasteurization
- Flash pasteurization
- Microwave chemistry
- Sterilization
- Food preservation
- Biofuel production

2.2.7 See also

- Specific absorption rate
- Electrosurgery, which requires direct joule heating of tissue, and thus directly transmitted high frequency currents

2.2.8 References

[1] Piyasena P; et al. (2003), "Radio frequency heating of foods: principles, applications and related properties—a review", *Crit Rev Food Sci Nutr.* **43** (6): 587–606, doi:10.1080/10408690390251129, PMID 14669879

[2]

[3] U.S. Patent 2,147,689. Method and apparatus for heating dielectric materials - J.G. Chafee

[4] "Diathermy", Collins English Dictionary - Complete & Unabridged 10th Edition. Retrieved August 29, 2013, from Dictionary.com website

2.2.9 External links

- Metaxas, A.C. (1996). *Foundations of Electroheat, A Unified Approach*. John Wiley & Sons. ISBN 0-471-95644-9.

- Metaxas, A.C., Meredith, R.J. (1983). *Industrial Microwave Heating (IEE Power Engineering Series)*. Institution of Engineering and Technology. ISBN 0-906048-89-3.

- U.S. Patent 2,147,689 – *Method and apparatus for heating dielectric materials*

2.3 Electromagnetic hypersensitivity

Electromagnetic hypersensitivity (**EHS**) is characterized by a group of symptoms purportedly caused by exposure to electromagnetic fields.[1] A more specific term used in medical literature is **idiopathic environmental intolerance attributed to electromagnetic fields** (**IEI-EMF**). Other terms for IEI-EMF include **electrohypersensitivity**, **electro-sensitivity**, and **electrical sensitivity** (**ES**). Idiopathic refers to the fact that the cause is unknown.

Although the thermal effects of electromagnetic fields on the body are established and used (e.g. diathermy), those who are self-described with electromagnetic hypersensitivity report responding to non-ionizing electromagnetic fields (or electromagnetic radiation) at intensities well below the maximum levels permitted by international radiation safety standards.

The reported symptoms of EHS include headache, fatigue, stress, sleep disturbances, skin symptoms like prickling, burning sensations and rashes, pain and ache in muscles and many other health problems. Whatever their cause, EHS symptoms are a real and sometimes disabling problem for the affected person. However, there is no scientific basis to link EHS symptoms to electromagnetic field exposure.[2]

The majority of provocation trials to date have found that self-described sufferers of electromagnetic hypersensitivity are unable to distinguish between exposure and non-exposure to electromagnetic fields,[3][4] and it is not recognized as a medical condition by the medical or scientific communities. Since a systematic review in 2005 showing no convincing scientific evidence for its being caused by electromagnetic fields,[3] several double-blind experiments have been published, each of which has suggested that people who report electromagnetic hypersensitivity are unable to detect the presence of electromagnetic fields and are as likely to report ill health following a sham exposure as they are following exposure to genuine electromagnetic fields, suggesting the cause to be the nocebo effect.[5][6][7]

2.3.1 Signs and symptoms

A 2001 survey found that people related their symptoms most frequently to mobile phone base stations (74%), followed by mobile phones (36%), cordless phones (29%), and power lines (27%). The survey was not designed to find any causal connection between electromagnetic field exposure and ill health.[8]

A report from the UK Health Protection Agency said that self-described "electrical sensitivity" sufferers have symptoms that can be grouped into two broad categories: facial skin symptoms and more general, non-specific symptoms across a range of body systems. The facial skin symptoms and their attribution to visual display units was mostly a Nordic phenomenon. The report pointed out that it did not "imply the acceptance of a causal relationship between symptoms and attributed exposure".[9]

Recently a smaller group of people in Europe and in the USA have reported general and severe symptoms such as headache, fatigue, tinnitus, dizziness, memory deficits, irregular heart beat, and whole-body skin symptoms.[10] A 2005 Health Protection Agency report noted an overlap in many peoples symptoms with other syndromes known as symptom-based conditions, functional somatic syndromes, and IEI (idiopathic environmental intolerance).[9] Levitt proposed ties between electromagnetic fields and some of these 20th-century conditions, including chronic fatigue syndrome, Gulf War syndrome, and autism.[11]

Those reporting electromagnetic hypersensitivity will usually describe different levels of susceptibility to electric fields, magnetic fields, and various frequencies of electromagnetic waves (including fluorescent and low-energy lights, and microwaves from mobile, cordless/portable phones), and WiFi with no consistency in the severity of symptoms between sufferers.[12] Other surveys of electromagnetic hypersensitivity sufferers have not been able to find any consistent pattern to these symptoms.[8][13] Instead, symptoms reflecting almost every part of the body have been attributed to electromagnetic field exposure.

A minority of people who report electromagnetic hypersensitivity claim to be severely affected by it. For instance, one survey has estimated that approximately 10% of electromagnetic hypersensitivity sufferers in Sweden were on sick leave or have taken early retirement or a disability pension, compared to 5% of the general population,[13] while a second survey has reported that of 3046 people who experienced 'annoyance' from electrical equipment, 340 (11%) reported 'much' annoyance.[14] For those who report being severely affected, their symptoms can have a significant impact on their quality of life; with sufferers reporting physical, mental and social impairment and psychological distress.[8]

2.3.2 Cause

World Health Organization

Following a study conducted in 2005, the World Health Organization (WHO) concluded that:

> EHS is characterized by a variety of non-specific symptoms that differ from individual to individual. The symptoms are certainly real and can vary widely in their severity. Whatever its cause, EHS can be a disabling problem for the affected individual. EHS has no clear diagnostic criteria and there is no scientific basis to link EHS symptoms to EMF exposure. Further, EHS is not a medical diagnosis, nor is it clear that it represents a single medical problem.[1]

Studies

Most blinded conscious provocation studies have failed to show a correlation between exposure and symptoms, leading to the suggestion that psychological mechanisms may play a role in causing or exacerbating EHS symptoms. In 2010 Rubin et al. published a follow-up to their 2005 review, bringing the totals to 46 double-blind experiments and 1175 individuals with self-diagnosed hypersensitivity.[3][15] Both reviews claimed that "no robust evidence could be found" to support the hypothesis that electromagnetic exposure causes EHS, as have other studies.[5][6] They also concluded that the studies supported the role of the nocebo effect in triggering acute symptoms in those with EHS, although it has been argued that this deduction cannot be made from observational studies,[4] and reports of children exhibiting the symptoms suggest that the nocebo effect may be unlikely in these cases.[16] The Essex provocation study of 2007 received some criticism for its methodology and analysis. In their response the authors noted that their study says nothing about long-term effects, but that those affected often claim to respond to the fields within a few minutes.[17]

Some psychologists have suggested that severely affected EHS people who claim that they are unable to live in a wireless society are, like hermits of ancient times, escaping from the pressures of modern life.[18] In addition, scare stories in the media seem capable of increasing the likelihood of the symptoms ascribed to electromagnetic exposure,[19] although another study questioned the validity of this argument on the grounds that ants' locomotion also shows adverse effects under electromagnetic exposure.[20]

On the other hand, a few provocation studies have claimed some correlation between exposure and symptoms, where subjects with self-diagnosed EHS have been screened for the frequency and type of exposure to which they are most sensitive.[21][22][23] It has been argued that these positive studies suggest that frequency or non-linear effects, proposed in 1979, rather than intensity, are relevant, although this has been disputed.[24][25][26] Provocation tests in 1981 suggested a link between use of video terminals and skin rash in sensitive individuals.[27] Some studies have suggested neurophysiological differences between sensitive individuals and controls. This may reflect either a psychophysiological stress response to participating in the study or a more general imbalance in autonomic nervous system regulation.[7][28][29][30][31] Although effects have been shown in some tests of effects on sleep, there are problems of intra-individual reproducibility.[32] Remediation studies suggest that removal of an electromagnetic environment may remove symptoms for computer workers and phantom limb pain.[33][34]

Some other types of studies suggest evidence for symptoms at non-thermal levels of electromagnetic exposure. A review in 2010 of ten studies on neurobehavioral and cancer outcomes near cell phone base stations found eight with increased prevalence, including sleep disturbance and headaches.[35] Since 1962 the microwave auditory effect or tinnitus has been shown from radio frequency exposure at levels below significant heating.[36][37] Studies during the 1960s, among workers in the USSR and Poland with occupational electromagnetic exposure, claimed to show a set of symptoms called the 'microwave syndrome'.[38][39][40][41] Since 2006 the World Health Organization has reported transient symptoms, such as vertigo, nausea, metallic taste and phosphenes, among workers moving through strong magnetic fields near MRI scanners at non-thermal levels.[42] This has been recognised in the ICNIRP Guidelines of 2010 and 2014,[43][44] and by the European Directive of 2013 under direct biophysical non-thermal undesired or unexpected health effects, such as stimulation of nerves or sensory organs creating temporary annoyance or affecting cognition or other brain functions.[45] There are some indications of a small subgroup of hyper-sensitive individuals.[46] Other areas under study include sensitivity shown through subliminal or autonomic effects as well as conscious effects. These include increased rates of stroke during geomagnetic events,[47][48] aurora sensitivity,[49] and cardiovascular changes or muscular excitation.[50][51][52] These effects do not necessarily relate to conscious sensitivity.

Other studies on sensitivity have looked at therapeutic procedures using non-thermal electromagnetic exposure,[53] genetic factors,[54] an alteration in mast cells, oxidative stress, protein expression and voltage-gated calcium channels.[55][56][57][58] Mercury release from dental amalgam and heavy metal toxicity have also been implicated in exposure effects and symptoms.[59] Another line of study has been the nature of hyper-sensitivity or intolerance and the range of environmental exposures which may be related to it. Some 80% of people with self-diagnosed electromagnetic intolerance also claim intolerance to low levels of

chemical exposure.[60][61]

In 2002, some controversy over the causal relationship was demonstrated by the Freiburger Appeal, a petition originated by the German environmental medical lobby group IGUMED, which stated that "we can see a clear temporal and spatial correlation between the appearance of [certain] disease and exposure to pulsed high-frequency microwave radiation", and demanding radical restrictions on mobile phone use.[62] To address some of these concerns, and others, Hocking advised in a 2006 WHO proceedings that the test type and duration should be tailored to the individual, and that washout times are needed to prevent a carry-over effect of previous exposure.[63] However, in 2005 the World Health Organization concluded that there is no known scientific basis for the belief that electromagnetic hypersensitivity is caused by exposure to an electromagnetic field.[1]

2.3.3 Diagnosis

Electromagnetic hypersensitivity is not currently an accepted diagnosis. At present, there are no accepted research criteria other than 'self-reported symptoms', and for clinicians there is no case definition or clinical practice guideline. There is no specific test that can identify sufferers, as symptoms other than skin disorders tend to be subjective or non-specific. It is important firstly to exclude all other possible causes of the symptoms. Researchers and the WHO have stressed the need for a careful investigation. For some, complaints of electromagnetic hypersensitivity may mask organic or psychiatric illness and requires both a thorough medical evaluation to identify and treat any specific conditions that may be responsible for the symptoms, and a psychological evaluation to identify alternative psychiatric/psychological conditions that may be responsible or contribute to the symptoms.[1][64]

A WHO factsheet also recommends an assessment of the workplace and home for factors that might contribute to the presented symptoms. These could include indoor air pollution, excessive noise, poor lighting (flickering light) or ergonomic factors. They also point out that "[some] studies suggest that certain physiological responses of [electromagnetic hypersensitivity] individuals tend to be outside the normal range. In particular, hyper reactivity in the central nervous system and imbalance in the autonomic nervous system need to be followed up in clinical investigations and the results for the individuals taken as input for possible treatment."[1]

2.3.4 Management

For individuals reporting electromagnetic hypersensitivity with long lasting symptoms and severe handicaps, treatment therapy should be directed principally at reducing symptoms and functional handicaps. This should be done in close co-operation with a qualified medical specialist to address the symptoms and a hygienist (to identify and, if necessary, control factors in the environment that have adverse health effects of relevance to the patient).[1]

Those who feel they are sensitive to electromagnetic fields generally try to reduce their exposure to electromagnetic sources as much as is practical. Complete avoidance of electromagnetic fields presents major practical difficulties in modern society. Methods often employed by sufferers include: avoiding sources of exposure; disconnecting or removing electrical devices; shielding or screening of self or residence; medication; and complementary and alternative therapy.[8]

The UK Health Protection Agency reviewed treatments for electromagnetic hypersensitivity, and success was reported with "neutralizing chemical dilution, antioxidant treatment, Cognitive Behavioural Therapy, Acupuncture and Shiatsu".[9] It was noted that:

> The studies reviewed suffer from a combination of the small numbers of subjects included and the potential variation both within and between study populations. Little information is given as to the attributed exposures of the subjects. These factors limit their general applicability outside the immediate study group. For those studies where detail was available, only two were placebo controlled [Acupunture and nutrition intervention].

It was also noted in the review that success may have more to do with offering a caring environment as opposed to a specific treatment.

A 2006 systematic review identified nine clinical trials testing different treatments for ES:[65] four studies tested cognitive behavioural therapy, two tested visual display unit filters, one tested a device emitting 'shielding' electromagnetic fields, one tested acupuncture, and one tested daily intake of tablets containing vitamin C, vitamin E, and selenium. The authors of the review concluded that:

> The evidence base concerning treatment options for electromagnetic hypersensitivity is limited and more research is needed before any definitive clinical recommendations can be made. However, the best evidence currently available suggests that cognitive behavioural therapy is effective for patients who report being hypersensitive to weak electromagnetic fields.

Some Americans with the condition have moved to the United States National Radio Quiet Zone where wireless is restricted.[66][67][68] Others have sought refuge by living off the grid.[69]

2.3.5   Prevalence

The prevalence of claimed electromagnetic hypersensitivity has been estimated as being between a few cases per million to 5% of the population depending on the location and definition of the condition.

In 2002, a questionnaire survey of 2,072 people in California found that the prevalence of self-reported electromagnetic hypersensitivity within the sample group was 3% (95% CI 2.8–3.68%), with electromagnetic hypersensitivity being defined as "being allergic or very sensitive to getting near electrical appliances, computers, or power lines" (response rate 58.3%).[70]

A similar questionnaire survey from the same year in Stockholm County (Sweden), found a 1.5% prevalence of self-reported electromagnetic hypersensitivity within the sample group, with electromagnetic hypersensitivity being defined as "hypersensitivity or allergy to electric or magnetic fields" (response rate 73%).[13]

A 2004 survey in Switzerland found a 5% prevalence of claimed electromagnetic hypersensitivity in the sample group of 2,048.[71]

In 2007, a UK survey aimed at a randomly selected group of 20,000 people found a prevalence of 4% for symptoms self-attributed to electromagnetic exposure.[72]

A group of scientists also attempted to estimate the number of people reporting "subjective symptoms" from electromagnetic fields for the European Commission.[73] In the words of a HPA review, they concluded that "the differences in prevalence were at least partly due to the differences in available information and media attention around electromagnetic hypersensitivity that exist in different countries. Similar views have been expressed by other commentators."[9]

2.3.6   Conferences

In 2004, the World Health Organization (WHO) conducted a workshop on electromagnetic hypersensitivity.[74] The aim of the conference was to review the current state of knowledge and opinions of the conference participants and propose ways forward on this issue. The meeting was conducted by the WHO International EMF Project as part of the scientific review process to determine biological and health effects from exposure to EMF. The purpose of these workshops is to bring together expert scientists so that established health effects and gaps in knowledge requiring further research can be identified. EHS has been a particularly contentious issue for a number of years.

2.3.7 In popular culture

In the AMC television program Better Call Saul, Saul's brother Chuck, played by Michael McKean, has become semi-reclusive and believes he suffers from electromagnetic hypersensitivity. Later in the series, Chuck's disease is shown to be psychosomatic.[75][76]

2.3.8 See also

- Arthur Firstenberg

- Electromagnetic radiation and health

- List of questionable diseases

- Mobile phone radiation and health

- TCO Certification for CRT monitor emission

- Wireless electronic devices and health

- Wi-Fi Safety

2.3.9 References

[1] "Electromagnetic fields and public health: Electromagnetic Hypersensitivity". *WHO Factsheet 296*. World Health Organisation (WHO). December 2005. Retrieved 2012-10-24.

[2] http://www.who.int/peh-emf/publications/facts/fs296/en/

[3] Rubin GJ, Das Munshi J, Wessely S (2005). "Electromagnetic hypersensitivity: a systematic review of provocation studies". *Psychosom Med* **67** (2): 224–232. doi:10.1097/01.psy.0000155664.13300.64. PMID 15784787.

[4] Röösli M (2008). "Radiofrequency electromagnetic field exposure and non-specific symptoms of ill health: a systematic review". *Environ. Res.* **107** (2): 277–287. doi:10.1016/j.envres.2008.02.003. PMID 18359015.

[5] Sabine Regel, Sonja Negovetic, Martin Röösli, Veronica Berdiñas, Jürgen Schuderer, Anke Huss, Urs Lott, Niels Kuster, Peter Achermann (August 2006). "UMTS Base Station-like Exposure, Well-Being, and Cognitive Performance". *Environ Health Perspect* **114** (8): 1270–5. doi:10.1289/ehp.8934. PMC 1552030. PMID 16882538.

[6] J Rubin, G Hahn, BS Everitt, AJ Clear, Simon Wessely (2006). "Are some people sensitive to mobile phone signals? Within participants double blind randomised provocation study". *British Medical Journal* **332** (7546): 886–889.

doi:10.1136/bmj.38765.519850.55. PMC 1440612. PMID 16520326.

[7] Wilén J, Johansson A, Kalezic N, Lyskov E, Sandström M (2006). "Psychophysiological tests and provocation of subjects with mobile phone related symptoms". *Bioelectromagnetics*. **27** (3): 204–214. doi:10.1002/bem.20195. PMID 16304699.

[8] Röösli, Martin; M Moser; Y Baldinini; M Meier; C Braun-Fahrländer (February 2004). "Symptoms of ill health ascribed to electromagnetic field exposure – a questionnaire survey". *Int J Hyg Environ Health* **207** (2): 141–50. doi:10.1078/1438-4639-00269. PMID 15031956.

[9] "Definition, epidemiology and management of electrical sensitivity", Irvine, N, Report for the Radiation Protection Division of the UK Health Protection Agency, HPA-RPD-010, 2005

[10] Sage, Cindy. "Microwave And Radiofrequency Radiation Exposure: A Growing Environmental Health Crisis?". San Francisco Medical Society web page. Retrieved 2008-05-31.

[11] Levitt, B. Blake (1995). *Electromagnetic Fields*. San Diego: Harcourt Brace & Company. pp. 181–218.

[12] Philips, Alasdair and Jean (2003–2011). Electromagnetic hypersensitivity (EHS) (in 8 sections)

[13] Hillert, L; N Berglind; BB Arnetz; T Bellander (February 2002). "Prevalence of self-reported hypersensitivity to electric or magnetic fields in a population-based questionnaire survey". *Scand J Work Environ Health* **28** (1): 33–41. doi:10.5271/sjweh.644. PMID 11871850.

[14] Carlsson, F; Karlson, B; Ørbæk, P; Österberg, K; Östergren, PO (July 2005). "Prevalence of annoyance attributed to electrical equipment and smells in a Swedish population, and relationship with subjective health and daily functioning". *Public Health* **119** (7): 568–77. doi:10.1016/j.puhe.2004.07.011. PMID 15925670.

[15] James Rubin, Rosa Nieto-Hernandez, Simon Wessely (2010). "Idiopathic Environmental Intolerance Attributed to Electromagnetic Fields". *Bioelectromagnetics* **31** (1): 1–11. doi:10.1002/bem.20536. PMID 19681059.

[16] Lamech F (2014). "Self-Reporting of Symptom Development From Exposure to Radiofrequency Fields of Wireless Smart Meters in Victoria, Australia: A Case Series". *Altern Ther Health Med.* **20** (6): 28–39. PMID 25478801.

[17] Cohen, A; Carlo, G; Davidson, A; White, M; Geoghan, C; Goldsworthy, A; Johansson, O; Maisch, D; O'Connor, E (2008-02-01). "Sensitivity to Mobile Phone Base Station Signals". *Environmental Health Perspectives* **116** (2): A63–4; author reply A64–5. doi:10.1289/ehp.10870. PMC 2235218. PMID 18288297.

[18] Boyd I, Rubin G, Wessely S (2012). "Taking refuge from modernity: 21st century hermits". *J R Soc Med.* **105** (12): 523–9. doi:10.1258/jrsm.2012.120060. PMID 23288087.

[19] Witthoft M, Rubin GJ (2013). "Are media warnings about the adverse health effects of modern life self-fulfilling? An experimental study on idiopathic environmental intolerance attributed to electromagnetic fields (IEI-EMF)". *J Psychosom Res.* **74** (3): 206–212. doi:10.1016/j.jpsychores.2012.12.002. PMID 23438710.

[20] Cammaerts MC, Johansson O (2014). "Ants can be used as bio-indicators to reveal biological effects of electromagnetic waves from some wireless apparatus". *Electromagn Biol Med.* **33** (4): 282–8. doi:10.3109/15368378.2013.817336. PMID 23977878.

[21] Rea W, Pan Y, Yenyves E, Sujisawa I, Suyama H, Samadi N, Ross G (1991). "Electromagnetic field sensitivity". *J Bioelectricity* **10** (1-2): 241–6.

[22] McCarty DE, Carrubba S, Chesson Jr. AL, Frilot II C, Gonzalez-Toledo E, Marino AA (2011). "Electromagnetic Hypersensitivity: evidence for a novel neurological syndrome". *Int J Neurosci.* **121** (12): 670–676. doi:10.3109/00207454.2011.608139. PMID 21793784.

[23] Köteles F, Szemerszky R, Gubányi M, Körmendi J, Szekrényesi C, Lloyd R, Molnár L, Drozdovszky O, Bárdos G (2013). "Idiopathic environmental intolerance attributed to electromagnetic fields (IEI-EMF) and electrosensibility (ES) - are they connected?". *Int J Hyg Environ Health.* **216** (3): 362–70. doi:10.1016/j.ijheh.2012.05.007. PMID 22698789.

[24] Adey WR (1979). "Neurophysiologic effects of radiofrequency and microwave radiation". *Bull NY Acad Med.* **55** (11): 1079–93. PMID 295243.

[25] Rubin GJ, Cleare AJ, Wessely S (2012). "Letter to the editor: electromagnetic hypersensitivity; author reply; discussion". *Int J Neurosci.* **122** (7): 401–4. doi:10.3109/00207454.2011.648763. PMID 22176592.

[26] Marino AA, = AA (2013). "Electromagnetic hypersensitivity syndrome revisited again". *Int J Neurosci.* **123** (8): 593–4. doi:10.3109/00207454.2013.775575. PMID 23410192.

[27] Lindén V, Rolfsen S (1981). "Video computer terminals and occupational dermatitis". *Scand J Work Environ Health.* **7** (1): 62–4. doi:10.5271/sjweh.2571. PMID 6458886.

[28] Lyskov E, Sandström M, Hansson Mild K (November 2001). "Neurophysiological study of patients with perceived 'electrical hypersensitivity'". *Int J Psychophysiol* **42** (3): 233–41. doi:10.1016/S0167-8760(01)00141-6. PMID 11812390.

[29] Sandström M, Lyskov E, Berglund A, Medvedev S, Mild KH (January 1997). "Neurophysiological effects of flickering light in patients with perceived electrical hypersensitivity". *J. Occup. Environ. Med.* **39** (1): 15–22. doi:10.1097/00043764-199701000-00006. PMID 9029427.

[30] Eltiti S, Wallace D, Ridgewell A, et al. (November 2007). "Does Short-Term Exposure to Mobile Phone Base Station Signals Increase Symptoms in Individuals Who Report Sensitivity to Electromagnetic Fields? A Double-Blind Randomized Provocation Study". *Environ. Health Perspect.* **115** (11): 1603–8. doi:10.1289/ehp.10286. PMC 2072835. PMID 18007992.

[31] Landgrebe M, Hauser S, Langguth B, Frick U, Hajak G, Eichhammer P (March 2007). "Altered cortical excitability in subjectively electrosensitive patients: results of a pilot study". *J Psychosom Res* **62** (3): 283–8. doi:10.1016/j.jpsychores.2006.11.007. PMID 17324677.

[32] Lustenberger C, Murbach M, Tüshaus L, Wehrle F, Kuster N, Achermann P, Huber R (2015). "Inter-individual and intra-individual variation of the effects of pulsed RF EMF exposure on the human sleep EEG". *Bioelectromagnetics.* doi:10.1002/bem.21893. PMID 25690404.

[33] Hagström M, Auranen J, Johansson O, Ekman R (2012). "Reducing electromagnetic irradiation and fields alleviates experienced health hazards of VDU work". *Pathophysiology.* **19** (2): 81–7. doi:10.1016/j.pathophys.2012.01.005. PMID 22364840.

[34] Fisher K, Oliver S, Sedki I, Hanspal R (Feb 2015). "The effect of electromagnetic shielding on phantom limb pain: A placebo-controlled double-blind crossover trial". *Prosthet Orthot Int.* doi:10.1177/0309364614568409. PMID 25716957.

[35] Khurana VG, Hardell L, Everaert J, Bortkiewicz A, Carlberg M, Ahonen M. "Epidemiological evidence for a health risk from mobile phone base stations". *Int J Occup Environ Health.* **16** (3): 263–7. doi:10.1179/107735210799160192. PMID 20662418.

[36] Frey AH (1962). "Human auditory system response to modulated electromagnetic energy". *J Appl Physiol.* **17** (4): 689–92. PMID 13895081.

[37] Hutter HP, Moshammer H, Wallner P, Cartellieri M, Denk-Linnert DM, Katzinger M, Ehrenberger K, Kundi M (2010). "Tinnitus and mobile phone use". *Occup Environ Med.* **67** (12): 804–8. doi:10.1136/oem.2009.048116. PMID 20573849.

[38] Czerski P, Hornowski J, Szewczykowski J: (1964). "Przypadek choroby mikrofalowej [Polish,"A case of Microwave Sickness"]". *Medycyna Pracy[Occupational Medicine]* **15** (4): 251–3.

[39] Bergman W (1965), *The Effect of Microwaves on the Central Nervous System (trans. from German)* (PDF), Ford Motor Company, pp. 1–77

[40] Marha K, Musil J, Tuha H (1968). *Electromagneticke pole a zivotni prostredi ["Electromagnetic Fields and the Life Environment"],.* National Health Publishing, Prague; (trans.) San Francisco Press, 1971. pp. 1–138. ASIN B0006C5GBK.

[41] Novitskiy Yu I, Gordon ZV, Presman AS, Kholodov Yu A (1970). *Radio Frequencies and Microwaves, Magnetic and Electric Fields, Foundations of Space Biology and Medicine (Radiochastoty i mikrovolny. Magnitnyye i elektricheskiye polya, Osnovy Kosmicheskoy Biologii i Meditsin)*, **2** (1.1). NASA Technical Translation, NASA TTF-1A,021; Moscow, Academy of Sciences USSR. pp. 1–288.

[42] World Health Organization (2006). "Static Fields" (PDF). *Environmental Health Criteria Monograph* **232**: 1–369.

[43] International Commission on Non-Ionizing Radiation Protection (2010). "Guidelines for limiting exposure to time-varying electric and magnetic (1 Hz – 100 kHz)" (PDF). *Health Phys.* **99** (6): 818–36. doi:10.1097/HP.0b013e3181f06c86. PMID 21068601.

[44] International Commission on Non-Ionizing Radiation Protection (2014). "Guidelines for limiting exposure to electric fields induced by movement of the human body in a static magnetic field and by time-varying magnetic fields below 1 Hz" (PDF). *Health Phys.* **106** (3): 418–25. doi:10.1097/HP.0b013e31829e5580. PMID 25208018.

[45] European Union (2013). "Directive 2013/35/EU of the European Parliament and of the Council, of 26 June 2013, on the minimum health and safety requirements regarding the exposure of workers to the risks arising from physical agents (electromagnetic fields) and repealing Directive 2004/40/EC". *Official Journal of the European Union.* L.179: 1–21.

[46] Schaap K, Christopher-de Vries Y, Mason CK, de Vocht F, Portengen L, Kromhout H (2014). "Occupational exposure of healthcare and research staff to static magnetic stray fields from 1.5-7 Tesla MRI scanners is associated with reporting of transient symptoms" (PDF). *Occup Environ Med.* **71** (6): 423–9. doi:10.1136/oemed-2013-101890. PMC 4033112. PMID 24714654.

[47] Shaposhnikov D, Revich B, Gurfinkel Y, Naumova E (2014). "The influence of meteorological and geomagnetic factors on acute myocardial infarction and brain stroke in Moscow, Russia". *Int J Biometeorol.* **58** (5): 799–808. doi:10.1007/s00484-013-0660-0. PMID 23700198.

[48] Palmer SJ, Rycroft MJ, Cermack M (2006). "Solar and geomagnetic activity, extremely low frequency magnetic and electric fields and human health at the Earth's surface". *Surv Geophysics.* **27** (5): 557–95. doi:10.1007/s10712-006-9010-7.

[49] Chernouss S, Vinogradov A, Vlassova E (2001). "Geophysical Hazard for Human Health in the Circumpolar Auroral Belt: Evidence of a Relationship between Heart Rate Variation and Electromagnetic Disturbances". *Nat Hazards.* **23** (2-3): 121–35. doi:10.1023/A:1011108723374.

[50] Havas M, Marrongelle J (2013). "Replication of heart rate variability provocation study with 2.4 GHz cordless phone confirms original findings". *Electromagn Biol Med.* **32**

(2): 253–66. doi:10.3109/15368378.2013.776437. PMID 23675629.

[51] Huttunen P, Savinainen A, Hänninen O, Myllylä R (2011). "Involuntary human hand movements due to FM radio waves in a moving van". *Acta Physiol Hung.* **98** (2): 157–64. doi:10.1556/APhysiol.98.2011.2.7. PMID 21616774.

[52] Huttunen P, Hänninen O, Myllylä R (2009). "FM-radio and TV tower signals can cause spontaneous hand movements near moving RF reflector". *Pathophysiology* **16** (3): 201–4. doi:10.1016/j.pathophys.2009.01.002. PMID 19268549.

[53] Pilla AA (2013). "Nonthermal electromagnetic fields: from first messenger to therapeutic applications". *Electromagn Biol Med.* **32** (2): 123–36. doi:10.3109/15368378.2013.776335. PMID 23675615.

[54] De Luca C, Chung Sheun Thai J, Raskovic D, Cesareo E, Caccamo D, Trukhanov A, Korkina L. "Metabolic and genetic screening of electromagnetic hypersensitive subjects as a feasible tool for diagnostics and intervention". *Mediators Inflamm.* **2014** (924184): 1–14. doi:10.1155/2014/924184. PMC 4000647. PMID 24812443.

[55] Johansson O, Gangi S, Liang Y, Yoshimura K, Jing C, Liu PY (2001). "Cutaneous mast cells are altered in normal healthy volunteers sitting in front of ordinary TVs/PCs - results from open-field provocation experiments". *J Cutan Pathol.* **28** (10): 513–9. doi:10.1034/j.1600-0560.2001.281004.x. PMID 11737520.

[56] Ghazizadeh V, Nazıroğlu M (2014). "Electromagnetic radiation (Wi-Fi) and epilepsy induce calcium entry and apoptosis through activation of TRPV1 channel in hippocampus and dorsal root ganglion of rats". *Metab Brain Dis.* **29** (3): 787–99. doi:10.1007/s11011-014-9549-9. PMID 24792079.

[57] Lu Y, He M, Zhang Y, Xu S, Zhang L, He Y, Chen C, Liu C, Pi H, Yu Z, Zhou Z (2014). "Differential Pro-Inflammatory Responses of Astrocytes and Microglia Involve STAT3 Activation in Response to 1800 MHz Radiofrequency Fields". *PLoS One.* **9** (10): e108318. doi:10.1371/journal.pone.0108318. PMID 25275372.

[58] Pall ML (2013). "Electromagnetic fields act via activation of voltage-gated calcium channels to produce beneficial or adverse effects". *J Cell Mol Med.* **17** (8): 958–65. doi:10.1111/jcmm.12088. PMID 23802593.

[59] Mortazavi SM, Daiee E, Yazdi A, Khiabani K, Kavousi A, Vazirinejad R, Behnejad B, Ghasemi M, Mood MB (2008). "Mercury release from dental amalgam restorations after magnetic resonance imaging and following mobile phone use". *Pak J Biol Sci.* **11** (8): 1142–6. doi:10.3923/pjbs.2008.1142.1146. PMID 18819554.

[60] Nordin S, Neely G, Olsson D, Sandström M (2014). "Odor and Noise Intolerance in Persons with Self-Reported Electromagnetic Hypersensitivity". *Int J Environ Res Public*

Health. **11** (9): 8794–8805. doi:10.3390/ijerph110908794. PMID 25166918.

[61] Genuis SJ, Lipp CT (2012). "Electromagnetic hypersensitivity: fact or fiction?". *Sci Total Environ.* **414**: 103–112. doi:10.1016/j.scitotenv.2011.11.008. PMID 22153604.

[62] Original text:"Betreff: Mobilfunk - Freiburger Appell" (in German). Interisziplinäre Gesellschaft für Umweltmedezin (IGUMED). 2002-10-09. Retrieved 2008-02-06. Translation: "Freiburger Appeal" (PDF). IGUMED. 2002-10-09. Archived from the original (PDF) on 2007-09-30. Retrieved 2008-02-06.

[63] Hocking, Bruce (2004-10-27). "A physician's approach to EMF sensitive patients" (PDF). *Proceedings; International Workshop on EMF Hypersensitivity; Prague, Czech Republic; October 25–27, 2004.* World Health Organization. Retrieved 2008-10-12.

[64] Rubin GJ, Cleare AJ, Wessely S (January 2008). "Psychological factors associated with self-reported sensitivity to mobile phones". *J Psychosom Res* **64** (1): 1–9; discussion 11–2. doi:10.1016/j.jpsychores.2007.05.006. PMID 18157992.

[65] Rubin GJ, Das Munshi J, Wessely S (2006). "A systematic review of treatments for electromagnetic hypersensitivity". *Psychother Psychosom* **75** (1): 12–8. doi:10.1159/000089222. PMID 16361870.

[66] O'Brien, Jane; Danzico, Matt (September 12, 2011). "'Wi-fi refugees' shelter in West Virginia mountains". BBC News. Retrieved September 13, 2011.

[67] Stromberg, Joseph (12 April 2013). "Green Bank, W.V., where the electrosensitive can escape the modern world. - Slate Magazine". *Slate.* Retrieved 14 April 2013.

[68] Gaynor, Michael (January 2015). "The Town Without Wi-Fi - Washingtonian". *Washingtonian.* Retrieved 12 January 2015.

[69] Johnson, Jeromy (January 2015). "EMF Refuge - Protect Your Family from EMF Pollution". Retrieved 12 January 2015.

[70] Levallois, P; R Neutra; G Lee; L Hristova (August 2002). "Study of self-reported hypersensitivity to electromagnetic fields in California". *Environ Health Perspect* **110** (Suppl 4): 619–23. doi:10.1289/ehp.02110s4619. PMC 1241215. PMID 12194896.

[71] Schreier N, Huss A, Röösli M (2006). "The prevalence of symptoms attributed to electromagnetic field exposure: a cross-sectional representative survey in Switzerland". *Soz Praventivmed* **51** (4): 202–9. doi:10.1007/s00038-006-5061-2. PMID 17193782.

[72] Eltiti S, Wallace D, Zougkou K, et al. (February 2007). "Development and evaluation of the electromagnetic hypersensitivity questionnaire". *Bioelectromagnetics* **28** (2): 137–51. doi:10.1002/bem.20279. PMID 17013888.

[73] Bergqvist, U; Vogel, E; Aringer, L; Cunningham, J; Gobba, F; Leitgeb, N; Miro, L; Neubauer, G; Ruppe, I; Vecchia, P; Wadman, C (1997). "Possible health implications of subjective symptoms and electromagnetic fields. A report prepared by a European group of experts for the European Commission, DG V". *Arbete och Hälsa* **19**.

[74] http://www.who.int/peh-emf/publications/reports/EHS_Proceedings_June2006.pdf

[75] Better Call Saul: is electromagnetic hypersensitivity a real health risk? , *The Guardian* (retrieved 7 Oct 2015).

[76] Better Call Saul's Michael McKean talks Chuck's condition and warns: do not know me too quickly, *Radio Times* (retrieved 7 Oct 2015).

2.4 Electromagnetic radiation

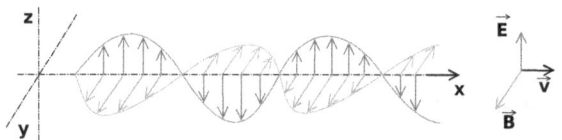

The electromagnetic waves that compose electromagnetic radiation can be imagined as a self-propagating transverse oscillating wave of electric and magnetic fields. This diagram shows a plane linearly polarized EMR wave propagating from left to right. The electric field is in a vertical plane and the magnetic field in a horizontal plane. The electric and magnetic fields in EMR waves are always in phase and at 90 degrees to each other.

Electromagnetic radiation (**EM radiation** or **EMR**) is the radiant energy released by certain electromagnetic processes. Visible light is one type of electromagnetic radiation, other familiar forms are invisible electromagnetic radiations such as radio waves, infrared light and X rays.

Classically, electromagnetic radiation consists of **electromagnetic waves**, which are synchronized oscillations of electric and magnetic fields that propagate at the speed of light through a vacuum. The oscillations of the two fields are perpendicular to each other and perpendicular to the direction of energy and wave propagation, forming a transverse wave. Electromagnetic waves can be characterized by either the frequency or wavelength of their oscillations to form the electromagnetic spectrum, which includes, in order of increasing frequency and decreasing wavelength: radio waves, microwaves, infrared radiation, visible light, ultraviolet radiation, X-rays and gamma rays.

Electromagnetic waves are produced whenever charged particles are accelerated, and these waves can subsequently interact with any charged particles. EM waves carry energy, momentum and angular momentum away from their source particle and can impart those quantities to matter with which they interact. Quanta of EM waves are called photons, which are massless, but they are still affected by gravity. Electromagnetic radiation is associated with those EM waves that are free to propagate themselves ("radiate") without the continuing influence of the moving charges that produced them, because they have achieved sufficient distance from those charges. Thus, EMR is sometimes referred to as the far field. In this jargon, the *near field* refers to EM fields near the charges and current that directly produced them, specifically, electromagnetic induction and electrostatic induction phenomena.

In the quantum theory of electromagnetism, EMR consists of photons, the elementary particles responsible for all electromagnetic interactions. Quantum effects provide additional sources of EMR, such as the transition of electrons to lower energy levels in an atom and black-body radiation. The energy of an individual photon is quantized and is greater for photons of higher frequency. This relationship is given by Planck's equation $E = h\nu$, where E is the energy per photon, ν is the frequency of the photon, and h is Planck's constant. A single gamma ray photon, for example, might carry ~100,000 times the energy of a single photon of visible light.

The effects of EMR upon biological systems (and also to many other chemical systems, under standard conditions) depend both upon the radiation's power and its frequency. For EMR of visible frequencies or lower (i.e., radio, microwave, infrared), the damage done to cells and other materials is determined mainly by power and caused primarily by heating effects from the combined energy transfer of many photons. By contrast, for ultraviolet and higher frequencies (i.e., X-rays and gamma rays), chemical materials and living cells can be further damaged beyond that done by simple heating, since individual photons of such high frequency have enough energy to cause direct molecular damage.

2.4.1 Physics

Theory

Main articles: Maxwell's equations and Near and far field

Maxwell's equations   James Clerk Maxwell first formally postulated *electromagnetic waves*. These were subsequently confirmed by Heinrich Hertz. Maxwell derived a wave form of the electric and magnetic equations, thus uncovering the wave-like nature of electric and magnetic fields and their symmetry. Because the speed of EM waves predicted by the wave equation coincided with the measured speed of

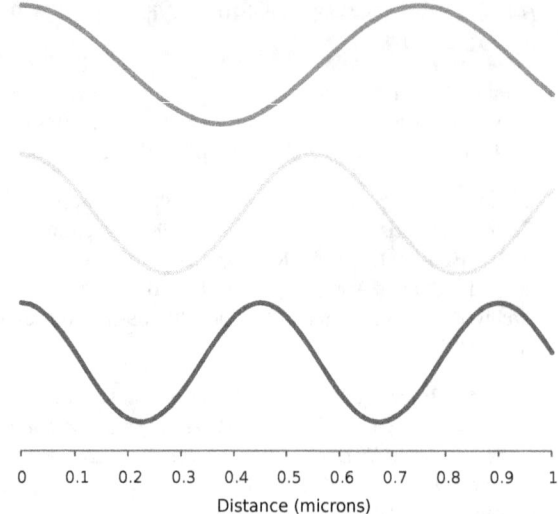

Shows the relative wavelengths of the electromagnetic waves of three different colours of light (blue, green, and red) with a distance scale in micrometers along the x-axis.

light, Maxwell concluded that light itself is an EM wave.

According to Maxwell's equations, a spatially varying electric field is always associated with a magnetic field that changes over time. Likewise, a spatially varying magnetic field is associated with specific changes over time in the electric field. In an electromagnetic wave, the changes in the electric field are always accompanied by a wave in the magnetic field in one direction, and vice versa. This relationship between the two occurs without either type field causing the other; rather, they occur together in the same way that time and space changes occur together and are interlinked in special relativity. In fact, magnetic fields may be viewed as relativistic distortions of electric fields, so the close relationship between space and time changes here is more than an analogy. Together, these fields form a propagating electromagnetic wave, which moves out into space and need never again affect the source. The distant EM field formed in this way by the acceleration of a charge carries energy with it that "radiates" away through space, hence the term.

Near and far fields   Main article: Liénard–Wiechert potential
Maxwell's equations established that some charges and currents ("sources") produce a local type of electromagnetic field near them that does *not* have the behaviour of EMR. Currents directly produce a magnetic field, but it is of a magnetic dipole type that dies out with distance from the current. In a similar manner, moving charges pushed apart in a conductor by a changing electrical potential (such as in an antenna) produce an electric dipole type electrical field,

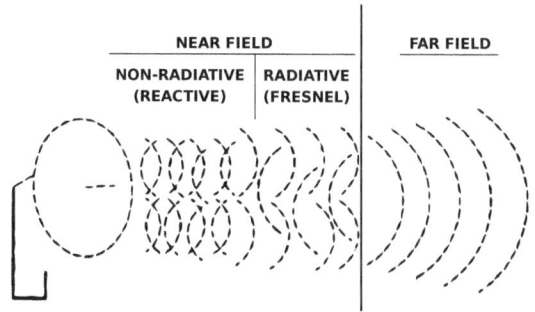

In electromagnetic radiation (such as microwaves from an antenna, shown here) the term applies only to the parts of the electromagnetic field that radiate into infinite space and decrease in intensity by an inverse-square law of power, so that the total radiation energy that crosses through an imaginary spherical surface is the same, no matter how far away from the antenna the spherical surface is drawn. Electromagnetic radiation thus includes the far field part of the electromagnetic field around a transmitter. A part of the "near-field" close to the transmitter, forms part of the changing electromagnetic field, but does not count as electromagnetic radiation.

but this also declines with distance. These fields make up the near-field near the EMR source. Neither of these behaviours are responsible for EM radiation. Instead, they cause electromagnetic field behaviour that only efficiently transfers power to a receiver very close to the source, such as the magnetic induction inside a transformer, or the feedback behaviour that happens close to the coil of a metal detector. Typically, near-fields have a powerful effect on their own sources, causing an increased "load" (decreased electrical reactance) in the source or transmitter, whenever energy is withdrawn from the EM field by a receiver. Otherwise, these fields do not "propagate" freely out into space, carrying their energy away without distance-limit, but rather oscillate, returning their energy to the transmitter if it is not received by a receiver.

By contrast, the EM far-field is composed of *radiation* that is free of the transmitter in the sense that (unlike the case in an electrical transformer) the transmitter requires the same power to send these changes in the fields out, whether the signal is immediately picked up or not. This distant part of the electromagnetic field *is* "electromagnetic radiation" (also called the far-field). The far-fields propagate (radiate) without allowing the transmitter to affect them. This causes them to be independent in the sense that their existence and their energy, after they have left the transmitter, is completely independent of both transmitter and receiver. Because such waves conserve the amount of energy they transmit through any spherical boundary surface drawn around their source, and because such surfaces have an area that is defined by the square of the distance from the source, the power of EM radiation always varies according to an inverse-square law. This is in contrast to dipole parts of the

EM field close to the source (the near-field), which varies in power according to an inverse cube power law, and thus does *not* transport a conserved amount of energy over distances, but instead fades with distance, with its energy (as noted) rapidly returning to the transmitter or absorbed by a nearby receiver (such as a transformer secondary coil).

The far-field (EMR) depends on a different mechanism for its production than the near-field, and upon different terms in Maxwell's equations. Whereas the magnetic part of the near-field is due to currents in the source, the magnetic field in EMR is due only to the local change in the electric field. In a similar way, while the electric field in the near-field is due directly to the charges and charge-separation in the source, the electric field in EMR is due to a change in the local magnetic field. Both processes for producing electric and magnetic EMR fields have a different dependence on distance than do near-field dipole electric and magnetic fields. That is why the EMR type of EM field becomes dominant in power "far" from sources. The term "far from sources" refers to how far from the source (moving at the speed of light) any portion of the outward-moving EM field is located, by the time that source currents are changed by the varying source potential, and the source has therefore begun to generate an outwardly moving EM field of a different phase.

A more compact view of EMR is that the far-field that composes EMR is generally that part of the EM field that has traveled sufficient distance from the source, that it has become completely disconnected from any feedback to the charges and currents that were originally responsible for it. Now independent of the source charges, the EM field, as it moves farther away, is dependent only upon the accelerations of the charges that produced it. It no longer has a strong connection to the direct fields of the charges, or to the velocity of the charges (currents).

In the Liénard–Wiechert potential formulation of the electric and magnetic fields due to motion of a single particle (according to Maxwell's equations), the terms associated with acceleration of the particle are those that are responsible for the part of the field that is regarded as electromagnetic radiation. By contrast, the term associated with the changing static electric field of the particle and the magnetic term that results from the particle's uniform velocity, are both associated with the electromagnetic near-field, and do not comprise EM radiation.

Properties

The physics of electromagnetic radiation is electrodynamics. Electromagnetism is the physical phenomenon associated with the theory of electrodynamics. Electric and magnetic fields obey the properties of

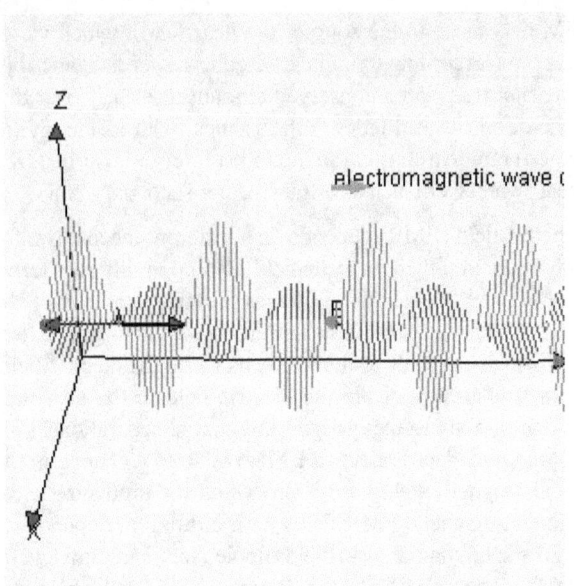

Electromagnetic waves can be imagined as a self-propagating transverse oscillating wave of electric and magnetic fields. This 3D animation shows a plane linearly polarized wave propagating from left to right. Note that the electric and magnetic fields in such a wave are in-phase with each other, reaching minima and maxima together

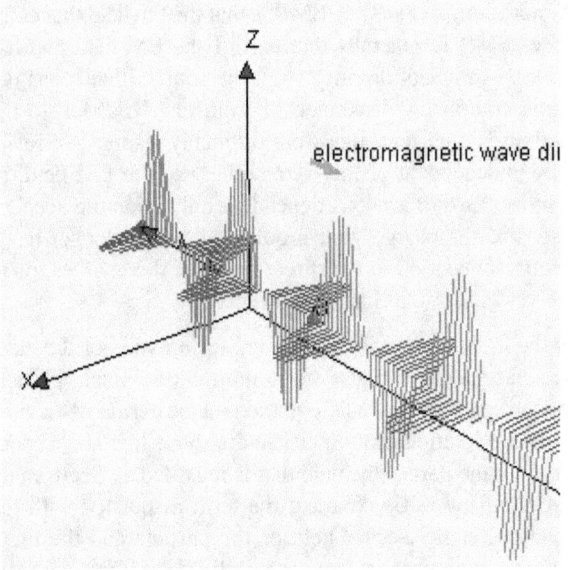

An alternate view of the wave shown above.

superposition. Thus, a field due to any particular particle or time-varying electric or magnetic field contributes to the fields present in the same space due to other causes. Further, as they are vector fields, all magnetic and electric field vectors add together according to vector addition. For example, in optics two or more coherent lightwaves may interact and by constructive or destructive interference

yield a resultant irradiance deviating from the sum of the component irradiances of the individual lightwaves.

Since light is an oscillation it is not affected by travelling through static electric or magnetic fields in a linear medium such as a vacuum. However, in nonlinear media, such as some crystals, interactions can occur between light and static electric and magnetic fields — these interactions include the Faraday effect and the Kerr effect.

In refraction, a wave crossing from one medium to another of different density alters its speed and direction upon entering the new medium. The ratio of the refractive indices of the media determines the degree of refraction, and is summarized by Snell's law. Light of composite wavelengths (natural sunlight) disperses into a visible spectrum passing through a prism, because of the wavelength-dependent refractive index of the prism material (dispersion); that is, each component wave within the composite light is bent a different amount.

EM radiation exhibits both wave properties and particle properties at the same time (see wave-particle duality). Both wave and particle characteristics have been confirmed in many experiments. Wave characteristics are more apparent when EM radiation is measured over relatively large timescales and over large distances while particle characteristics are more evident when measuring small timescales and distances. For example, when electromagnetic radiation is absorbed by matter, particle-like properties will be more obvious when the average number of photons in the cube of the relevant wavelength is much smaller than 1. It is not too difficult to experimentally observe non-uniform deposition of energy when light is absorbed, however this alone is not evidence of "particulate" behavior. Rather, it reflects the quantum nature of *matter*.[1] Demonstrating that the light itself is quantized, not merely its interaction with matter, is a more subtle affair.

Some experiments display both the wave and particle natures of electromagnetic waves, such as the self-interference of a single photon.[2] When a single photon is sent through an interferometer, it passes through both paths, interfering with itself, as waves do, yet is detected by a photomultiplier or other sensitive detector only once.

A quantum theory of the interaction between electromagnetic radiation and matter such as electrons is described by the theory of quantum electrodynamics.

Wave model

Electromagnetic radiation is a transverse wave, meaning that its oscillations are perpendicular to the direction of energy transfer and travel. The electric and magnetic parts of the field stand in a fixed ratio of strengths in order to satisfy the two Maxwell equations that specify how one is produced from the other. These **E** and **B** fields are also in phase, with both reaching maxima and minima at the same points in space (see illustrations). A common misconception is that the **E** and **B** fields in electromagnetic radiation are out of phase because a change in one produces the other, and this would produce a phase difference between them as sinusoidal functions (as indeed happens in electromagnetic induction, and in the near-field close to antennas). However, in the far-field EM radiation which is described by the two source-free Maxwell curl operator equations, a more correct description is that a time-change in one type of field is proportional to a space-change in the other. These derivatives require that the **E** and **B** fields in EMR are in-phase (see math section below).

An important aspect of light's nature is its frequency. The frequency of a wave is its rate of oscillation and is measured in hertz, the SI unit of frequency, where one hertz is equal to one oscillation per second. Light usually has multiple frequencies that sum to form the resultant wave. Different frequencies undergo different angles of refraction, a phenomenon known as dispersion.

A wave consists of successive troughs and crests, and the distance between two adjacent crests or troughs is called the wavelength. Waves of the electromagnetic spectrum vary in size, from very long radio waves the size of buildings to very short gamma rays smaller than atom nuclei. Frequency is inversely proportional to wavelength, according to the equation:

$$v = f\lambda$$

where v is the speed of the wave (c in a vacuum, or less in other media), f is the frequency and λ is the wavelength. As waves cross boundaries between different media, their speeds change but their frequencies remain constant.

Electromagnetic waves in free space must be solutions of Maxwell's electromagnetic wave equation. Two main classes of solutions are known, namely plane waves and spherical waves. The plane waves may be viewed as the limiting case of spherical waves at a very large (ideally infinite) distance from the source. Both types of waves can have a waveform which is an arbitrary time function (so long as it is sufficiently differentiable to conform to the wave equation). As with any time function, this can be decomposed by means of Fourier analysis into its frequency spectrum, or individual sinusoidal components, each of which contains a single frequency, amplitude and phase. Such a component wave is said to be *monochromatic*. A monochromatic electromagnetic wave can be characterized by its frequency or wavelength, its peak amplitude, its phase relative to some reference phase, its direction of propagation and its polar-

ization.

Interference is the superposition of two or more waves resulting in a new wave pattern. If the fields have components in the same direction, they constructively interfere, while opposite directions cause destructive interference. An example of interference caused by EMR is electromagnetic interference (EMI) or as it is more commonly known as, radio-frequency interference (RFI).

The energy in electromagnetic waves is sometimes called radiant energy.

Particle model and quantum theory

See also: Quantization (physics) and Quantum optics

An anomaly arose in the late 19th century involving a contradiction between the wave theory of light and measurements of the electromagnetic spectra that were being emitted by thermal radiators known as black bodies. Physicists struggled with this problem, which later became known as the ultraviolet catastrophe, unsuccessfully for many years. In 1900, Max Planck developed a new theory of black-body radiation that explained the observed spectrum. Planck's theory was based on the idea that black bodies emit light (and other electromagnetic radiation) only as discrete bundles or packets of energy. These packets were called quanta. Later, Albert Einstein proposed that light quanta be regarded as real particles. Later the particle of light was given the name photon, to correspond with other particles being described around this time, such as the electron and proton. A photon has an energy, E, proportional to its frequency, f, by

$$E = hf = \frac{hc}{\lambda}$$

where h is Planck's constant, λ is the wavelength and c is the speed of light. This is sometimes known as the Planck–Einstein equation.[3] In quantum theory (see first quantization) the energy of the photons is thus directly proportional to the frequency of the EMR wave.[4]

Likewise, the momentum p of a photon is also proportional to its frequency and inversely proportional to its wavelength:

$$p = \frac{E}{c} = \frac{hf}{c} = \frac{h}{\lambda}.$$

The source of Einstein's proposal that light was composed of particles (or could act as particles in some circumstances) was an experimental anomaly not explained by the wave theory: the photoelectric effect, in which light striking a metal surface ejected electrons from the surface, causing an electric current to flow across an applied voltage. Experimental measurements demonstrated that the energy of individual ejected electrons was proportional to the *frequency*, rather than the *intensity*, of the light. Furthermore, below a certain minimum frequency, which depended on the particular metal, no current would flow regardless of the intensity. These observations appeared to contradict the wave theory, and for years physicists tried in vain to find an explanation. In 1905, Einstein explained this puzzle by resurrecting the particle theory of light to explain the observed effect. Because of the preponderance of evidence in favor of the wave theory, however, Einstein's ideas were met initially with great skepticism among established physicists. Eventually Einstein's explanation was accepted as new particle-like behavior of light was observed, such as the Compton effect.

As a photon is absorbed by an atom, it excites the atom, elevating an electron to a higher energy level (one that is on average farther from the nucleus). When an electron in an excited molecule or atom descends to a lower energy level, it emits a photon of light at a frequency corresponding to the energy difference. Since the energy levels of electrons in atoms are discrete, each element and each molecule emits and absorbs its own characteristic frequencies. Immediate photon emission is called fluorescence, a type of photoluminescence. An example is visible light emitted from fluorescent paints, in response to ultraviolet (blacklight). Many other fluorescent emissions are known in spectral bands other than visible light. Delayed emission is called phosphorescence.

Wave–particle duality

The modern theory that explains the nature of light includes the notion of wave–particle duality. More generally, the theory states that everything has both a particle nature and a wave nature, and various experiments can be done to bring out one or the other. The particle nature is more easily discerned using an object with a large mass. A bold proposition by Louis de Broglie in 1924 led the scientific community to realize that electrons also exhibited wave–particle duality.

Wave and particle effects of electromagnetic radiation

Together, wave and particle effects fully explain the emission and absorption spectra of EM radiation. The matter-composition of the medium through which the light travels determines the nature of the absorption and emission spectrum. These bands correspond to the allowed energy levels in the atoms. Dark bands in the absorption spectrum are due to the atoms in an intervening medium between source and observer. The atoms absorb certain frequencies of the light between emitter and detector/eye, then emit them in all directions. A dark band appears to the detector, due to the radiation scattered out of the beam. For instance, dark bands in the light emitted by a distant star are due to the atoms in the star's atmosphere. A similar phenomenon occurs for emission, which is seen when an emitting gas glows due to excitation of the atoms from any mechanism, including heat. As electrons descend to lower energy levels, a spectrum is emitted that represents the jumps between the energy levels of the electrons, but lines are seen because again emission happens only at particular energies after excitation. An example is the emission spectrum of nebulae. Rapidly moving electrons are most sharply accelerated when they encounter a region of force, so they are responsible for producing much of the highest frequency electromagnetic radiation observed in nature.

These phenomena can aid various chemical determinations for the composition of gases lit from behind (absorption spectra) and for glowing gases (emission spectra). Spectroscopy (for example) determines what chemical elements comprise a particular star. Spectroscopy is also used in the determination of the distance of a star, using the red shift.

Propagation speed

Main article: Speed of light

Any electric charge that accelerates, or any changing magnetic field, produces electromagnetic radiation. Electromagnetic information about the charge travels at the speed of light. Accurate treatment thus incorporates a concept known as retarded time, which adds to the expressions for the electrodynamic electric field and magnetic field. These extra terms are responsible for electromagnetic radiation.

When any wire (or other conducting object such as an antenna) conducts alternating current, electromagnetic radiation is propagated at the same frequency as the current. In many such situations it is possible to identify an electrical dipole moment that arises from separation of charges due to the exciting electrical potential, and this dipole moment oscillates in time, as the charges move back and forth. This oscillation at a given frequency gives rise to changing electric and magnetic fields, which then set the electromagnetic radiation in motion.

At the quantum level, electromagnetic radiation is produced when the wavepacket of a charged particle oscillates or otherwise accelerates. Charged particles in a stationary state do not move, but a superposition of such states may result in a transition state that has an electric dipole moment that oscillates in time. This oscillating dipole moment is responsible for the phenomenon of radiative transition between quantum states of a charged particle. Such states occur (for example) in atoms when photons are radiated as the atom shifts from one stationary state to another.

As a wave, light is characterized by a velocity (the speed of light), wavelength, and frequency. As particles, light is a stream of photons. Each has an energy related to the frequency of the wave given by Planck's relation $E = hf$, where E is the energy of the photon, $h = 6.626 \times 10^{-34}$ J·s is Planck's constant, and f is the frequency of the wave.

One rule is obeyed regardless of circumstances: EM radiation in a vacuum travels at the speed of light, *relative to the observer*, regardless of the observer's velocity. (This observation led to Einstein's development of the theory of special relativity.)

In a medium (other than vacuum), velocity factor or refractive index are considered, depending on frequency and application. Both of these are ratios of the speed in a medium to speed in a vacuum.

Special theory of relativity

Main article: Special theory of relativity

By the late nineteenth century, various experimental anomalies could not be explained by the simple wave theory. One of these anomalies involved a controversy over the speed of light. The speed of light and other EMR predicted by Maxwell's equations did not appear unless the equations were modified in a way first suggested by FitzGerald and Lorentz (see history of special relativity), or else otherwise that speed would depend on the speed of observer relative to the "medium" (called luminiferous aether) which supposedly "carried" the electromagnetic wave (in a manner analogous to the way air carries sound waves). Experiments failed to find any observer effect. In 1905, Einstein proposed that space and time appeared to be velocity-changeable entities for light propagation and all other processes and laws. These changes accounted for the constancy of the speed of light and all electromagnetic radiation, from the viewpoints of all observers—even those in relative motion.

2.4.2 History of discovery

Electromagnetic radiation of wavelengths other than those of visible light were discovered in the early 19th century. The discovery of infrared radiation is ascribed to astronomer William Herschel, who published his results in 1800 before the Royal Society of London.[5] Herschel used a glass prism to refract light from the Sun and detected invisible rays that caused heating beyond the red part

of the spectrum, through an increase in the temperature recorded with a thermometer. These "calorific rays" were later termed infrared.

In 1801, German physicist Johann Wilhelm Ritter discovered ultraviolet in an experiment similar to Hershel's, using sunlight and a glass prism. Ritter noted that invisible rays near the violet edge of a solar spectrum dispersed by a triangular prism darkened silver chloride preparations more quickly than did the nearby violet light. Ritter's experiments were an early precursor to what would become photography. Ritter noted that the ultraviolet rays (which at first were called "chemical rays") were capable of causing chemical reactions.

In 1862-4 James Clerk Maxwell developed equations for the electromagnetic field which suggested that waves in the field would travel with a speed that was very close to the known speed of light. Maxwell therefore suggested that visible light (as well as invisible infrared and ultraviolet rays by inference) all consisted of propagating disturbances (or radiation) in the electromagnetic field. Radio waves were first produced deliberately by Heinrich Hertz in 1887, using electrical circuits calculated to produce oscillations at a much lower frequency than that of visible light, following recipes for producing oscillating charges and currents suggested by Maxwell's equations. Hertz also developed ways to detect these waves, and produced and characterized what were later termed radio waves and microwaves.[6]:286,7

Wilhelm Röntgen discovered and named X-rays. After experimenting with high voltages applied to an evacuated tube on 8 November 1895, he noticed a fluorescence on a nearby plate of coated glass. In one month, he discovered X-rays' main properties.[6]:307

The last portion of the EM spectrum to be discovered was associated with radioactivity. Henri Becquerel found that uranium salts caused fogging of an unexposed photographic plate through a covering paper in a manner similar to X-rays, and Marie Curie discovered that only certain elements gave off these rays of energy, soon discovering the intense radiation of radium. The radiation from pitchblende was differentiated into alpha rays (alpha particles) and beta rays (beta particles) by Ernest Rutherford through simple experimentation in 1899, but these proved to be charged particulate types of radiation. However, in 1900 the French scientist Paul Villard discovered a third neutrally charged and especially penetrating type of radiation from radium, and after he described it, Rutherford realized it must be yet a third type of radiation, which in 1903 Rutherford named gamma rays. In 1910 British physicist William Henry Bragg demonstrated that gamma rays are electromagnetic radiation, not particles, and in 1914 Rutherford and Edward Andrade measured their wavelengths, finding that they were similar to X-rays but with shorter wavelengths and higher frequency, although a 'cross-over' between X and gamma rays makes it possible to have X-rays with a higher energy (and hence shorter wavelength) than gamma rays and vice versa. The origin of the ray differentiates them, gamma rays tend to be a natural phenomena originating from the unstable nucleus of an atom and X-rays are electrically generated (and hence man-made) unless they are as a result of bremsstrahlung X-radiation caused by the interaction of fast moving particles (such as beta particles) colliding with certain materials, usually of higher atomic numbers.[6]:308,9

2.4.3 Electromagnetic spectrum

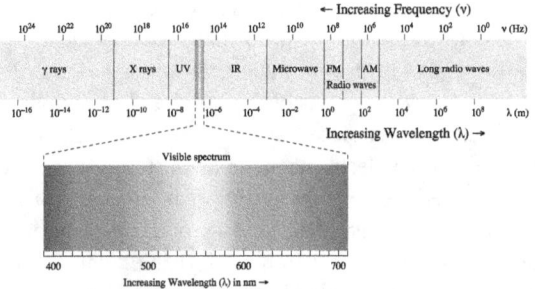

Electromagnetic spectrum with visible light highlighted

Main article: Electromagnetic spectrum

 EM radiation (the designation 'radiation' excludes static electric and magnetic and near fields) is classified by wavelength into radio, microwave, infrared, visible, ultraviolet, X-rays and gamma rays. Arbitrary electromagnetic waves can be expressed by Fourier analysis in terms of sinusoidal monochromatic waves, which in turn can each be classified into these regions of the EMR spectrum.

For certain classes of EM waves, the waveform is most usefully treated as *random*, and then spectral analysis must be done by slightly different mathematical techniques appropriate to random or stochastic processes. In such cases, the individual frequency components are represented in terms of their *power* content, and the phase information is not preserved. Such a representation is called the power spectral density of the random process. Random electromagnetic radiation requiring this kind of analysis is, for example, encountered in the interior of stars, and in certain other very wideband forms of radiation such as the Zero point wave field of the electromagnetic vacuum.

The behavior of EM radiation depends on its frequency. Lower frequencies have longer wavelengths, and higher frequencies have shorter wavelengths, and are associated with photons of higher energy. There is no fundamental limit known to these wavelengths or energies, at either end of the spectrum, although photons with energies near the Planck

CLASS	FREQUENCY	WAVELENGTH	ENERGY
γ	300 EHz	1 pm	1.24 MeV
HX	30 EHz	10 pm	124 keV
	3 EHz	100 pm	12.4 keV
SX	300 PHz	1 nm	1.24 keV
EUV	30 PHz	10 nm	124 eV
NUV	3 PHz	100 nm	12.4 eV
NIR	300 THz	1 μm	1.24 eV
MIR	30 THz	10 μm	124 meV
FIR	3 THz	100 μm	12.4 meV
	300 GHz	1 mm	1.24 meV
EHF	30 GHz	1 cm	124 μeV
SHF	3 GHz	1 dm	12.4 μeV
UHF	300 MHz	1 m	1.24 μeV
VHF	30 MHz	10 m	124 neV
HF	3 MHz	100 m	12.4 neV
MF	300 kHz	1 km	1.24 neV
LF	30 kHz	10 km	124 peV
VLF	3 kHz	100 km	12.4 peV
VF/ULF	300 Hz	1 Mm	1.24 peV
SLF	30 Hz	10 Mm	124 feV
ELF	3 Hz	100 Mm	12.4 feV

Legend:
γ = Gamma rays
HX = Hard X-rays
SX = Soft X-Rays
EUV = Extreme-ultraviolet
NUV = Near-ultraviolet
Visible light (colored bands)
NIR = Near-infrared
MIR = Mid-infrared
FIR = Far-infrared
EHF = Extremely high frequency (microwaves)
SHF = Super-high frequency (microwaves)
UHF = Ultrahigh frequency (radio waves)
VHF = Very high frequency (radio)
HF = High frequency (radio)
MF = Medium frequency (radio)
LF = Low frequency (radio)
VLF = Very low frequency (radio)
VF = Voice frequency
ULF = Ultra-low frequency (radio)
SLF = Super-low frequency (radio)
ELF = Extremely low frequency(radio)

energy or exceeding it (far too high to have ever been observed) will require new physical theories to describe.

Soundwaves are not electromagnetic radiation. At the lower end of the electromagnetic spectrum, about 20 Hz to about 20 kHz, are frequencies that might be considered in the audio range. However, electromagnetic waves cannot be directly perceived by human ears. Sound waves are instead the oscillating compression of molecules. To be heard, electromagnetic radiation must be converted to pressure waves of the fluid in which the ear is located (whether the fluid is air, water or something else).

Interactions as a function of frequency

When EM radiation interacts with matter, its behavior changes qualitatively as its frequency changes.

Radio and microwave   At radio and microwave frequencies, EMR interacts with matter largely as a bulk collection of charges which are spread out over large numbers of affected atoms. In electrical conductors, such induced bulk movement of charges (electric currents) results in absorption of the EMR, or else separations of charges that cause generation of new EMR (effective reflection of the EMR). An example is absorption or emission of radio waves by antennas, or absorption of microwaves by water or other molecules with an electric dipole moment, as for example inside a microwave oven. These interactions produce either electric currents or heat, or both. Infrared EMR interacts with dipoles present in single molecules, which change as atoms vibrate at the ends of a single chemical bond. For this reason, infrared is reflected by metals (as is most EMR into the ultraviolet) but is absorbed by a wide range of substances, causing them to increase in temperature as the vibrations dissipate as heat. In the same process, bulk substances radiate in the infrared spontaneously (see thermal radiation section below).

Visible light   As frequency increases into the visible range, photons of EMR have enough energy to change the bond structure of some individual molecules. It is not a coincidence that this happens in the "visible range," as the mechanism of vision involves the change in bonding of a single molecule (retinal) which absorbs light in the rhodopsin in the retina of the human eye. Photosynthesis becomes possible in this range as well, for similar reasons, as a single molecule of chlorophyll is excited by a single photon. Animals that detect infrared make use of small packets of water that change temperature, in an essentially thermal process that involves many photons (see infrared sensing in snakes). For this reason, infrared, microwaves and radio waves are thought to damage molecules and biological tissue only by bulk heating, not excitation from single photons of the radiation.

Visible light is able to affect a few molecules with single photons, but usually not in a permanent or damaging way, in the absence of power high enough to increase temperature to damaging levels. However, in plant tissues that conduct photosynthesis, carotenoids act to quench electronically excited chlorophyll produced by visible light in a process called non-photochemical quenching, in order to prevent reactions that would otherwise interfere with photosynthesis at high light levels. Limited evidence indicate that some reactive oxygen species are created by visible light in skin, and that these may have some role in photoaging, in the same manner as ultraviolet A.[7]

Ultraviolet   As frequency increases into the ultraviolet, photons now carry enough energy (about three electron volts or more) to excite certain doubly bonded molecules into permanent chemical rearrangement. In DNA, this causes lasting damage. DNA is also indirectly damaged by reactive oxygen species produced by ultraviolet A (UVA), which has energy too low to damage DNA directly. This is why ultraviolet at all wavelengths can damage DNA, and is capable of causing cancer, and (for UVB) skin burns (sunburn) that are far worse than would be produced by simple heating (temperature increase) effects. This property of causing molecular damage that is out of proportion to heating effects, is characteristic of all EMR with frequencies at the visible light range and above. These properties of high-frequency EMR are due to quantum effects that permanently damage materials and tissues at the molecular level.

At the higher end of the ultraviolet range, the energy of photons becomes large enough to impart enough energy to electrons to cause them to be liberated from the atom, in a process called photoionisation. The energy required for this is always larger than about 10 electron volts (eV) corresponding with wavelengths smaller than 124 nm (some sources suggest a more realistic cutoff of 33 eV, which is the energy required to ionize water). This high end of the ultraviolet spectrum with energies in the approximate ionization range, is sometimes called "extreme UV." Ionizing UV is strongly filtered by the Earth's atmosphere).

X-rays and gamma rays   Electromagnetic radiation composed of photons that carry minimum-ionization energy, or more, (which includes the entire spectrum with shorter wavelengths), is therefore termed ionizing radiation. (Many other kinds of ionizing radiation are made of non-EM particles). Electromagnetic-type ionizing radiation extends from the extreme ultraviolet to all higher frequencies and shorter wavelengths, which means that all X-rays and gamma rays qualify. These are capable of the most severe types of molecular damage, which can happen in biology to any type of biomolecule, including mutation and cancer, and often at great depths below the skin, since the higher end of the X-ray spectrum, and all of the gamma ray spectrum, penetrate matter. This type of damage causes these types of radiation to be especially carefully monitored, due to their hazard, even at comparatively low-energies, to all living organisms.

2.4.4 Atmosphere and magnetosphere

Main articles: ozone layer, shortwave radio, skywave and ionosphere
 Most UV and X-rays are blocked by absorption first from

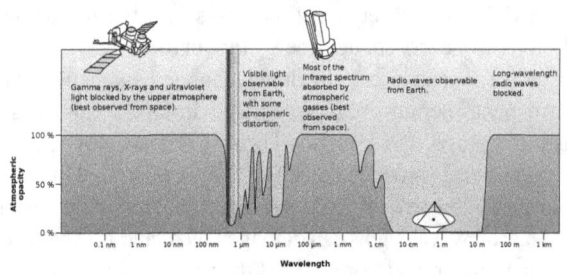

Rough plot of Earth's atmospheric absorption and scattering (or opacity) of various wavelengths of electromagnetic radiation

molecular nitrogen, and then (for wavelengths in the upper UV) from the electronic excitation of dioxygen and finally ozone at the mid-range of UV. Only 30% of the Sun's ultraviolet light reaches the ground, and almost all of this is well transmitted.

Visible light is well transmitted in air, as it is not energetic enough to excite nitrogen, oxygen, or ozone, but too energetic to excite molecular vibrational frequencies of water vapor.

Absorption bands in the infrared are due to modes of vibrational excitation in water vapor. However, at energies too low to excite water vapor, the atmosphere becomes transparent again, allowing free transmission of most microwave and radio waves.

Finally, at radio wavelengths longer than 10 meters or so (about 30 MHz), the air in the lower atmosphere remains transparent to radio, but plasma in certain layers of the ionosphere begins to interact with radio waves (see skywave). This property allows some longer wavelengths (100 meters or 3 MHz) to be reflected and results in shortwave radio beyond line-of-sight. However, certain ionospheric effects begin to block incoming radiowaves from space, when their frequency is less than about 10 MHz (wavelength longer than about 30 meters).

2.4.5 Types and sources, classed by spectral band

See electromagnetic spectrum

Radio waves

Main article: Radio waves

When radio waves impinge upon a conductor, they couple to the conductor, travel along it and induce an electric current on the conductor surface by moving the electrons of the conducting material in correlated bunches of charge. Such effects can cover macroscopic distances in conductors (including as radio antennas), since the wavelength of radiowaves is long.

Microwaves

Main article: Microwaves

Microwaves are a form of electromagnetic radiation with wavelengths ranging from as long as one meter to as short as one millimeter; with frequencies between 300 MHz (0.3 GHz) and 300 GHz.

Infrared

Main article: Infrared

Visible light

Main article: Light

Natural sources produce EM radiation across the spectrum. EM radiation with a wavelength between approximately 400 nm and 700 nm is directly detected by the human eye and perceived as visible light. Other wavelengths, especially nearby infrared (longer than 700 nm) and ultraviolet (shorter than 400 nm) are also sometimes referred to as light.

Ultraviolet

Main article: Ultraviolet

X-rays

Main article: X-rays

Gamma rays

Main article: Gamma rays

Thermal radiation and electromagnetic radiation as a form of heat

Main articles: Thermal radiation and Planck's law

The basic structure of matter involves charged particles bound together. When electromagnetic radiation impinges on matter, it causes the charged particles to oscillate and gain energy. The ultimate fate of this energy depends on the context. It could be immediately re-radiated and appear as scattered, reflected, or transmitted radiation. It may get dissipated into other microscopic motions within the matter, coming to thermal equilibrium and manifesting itself as thermal energy in the material. With a few exceptions related to high-energy photons (such as fluorescence, harmonic generation, photochemical reactions, the photovoltaic effect for ionizing radiations at far ultraviolet, X-ray and gamma radiation), absorbed electromagnetic radiation simply deposits its energy by heating the material. This happens for infrared, microwave and radio wave radiation. Intense radio waves can thermally burn living tissue and can cook food. In addition to infrared lasers, sufficiently intense visible and ultraviolet lasers can easily set paper afire.

Ionizing radiation creates high-speed electrons in a material and breaks chemical bonds, but after these electrons collide many times with other atoms eventually most of the energy becomes thermal energy all in a tiny fraction of a second. This process makes ionizing radiation far more dangerous per unit of energy than non-ionizing radiation. This caveat also applies to UV, even though almost all of it is not ionizing, because UV can damage molecules due to electronic excitation, which is far greater per unit energy than heating effects.

Infrared radiation in the spectral distribution of a black body is usually considered a form of heat, since it has an equivalent temperature and is associated with an entropy change per unit of thermal energy. However, "heat" is a technical term in physics and thermodynamics and is often confused with thermal energy. Any type of electromagnetic energy can be transformed into thermal energy in interaction with matter. Thus, *any* electromagnetic radiation can "heat" (in the sense of increase the thermal energy termperature of) a material, when it is absorbed.

The inverse or time-reversed process of absorption is thermal radiation. Much of the thermal energy in matter consists of random motion of charged particles, and this energy can be radiated away from the matter. The resulting radiation may subsequently be absorbed by another piece of matter, with the deposited energy heating the material.

The electromagnetic radiation in an opaque cavity at thermal equilibrium is effectively a form of thermal energy, having maximum radiation entropy.

2.4.6 Biological effects

Main articles: Electromagnetic radiation and health and Mobile phone radiation and health

Bioelectromagnetics is the study of the interactions and effects of EM radiation on living organisms. The effects of electromagnetic radiation upon living cells, including those in humans, depends upon the radiation's power and frequency. For low-frequency radiation (radio waves to visible light) the best-understood effects are those due to radiation power alone, acting through heating when radiation is absorbed. For these thermal effects, frequency is important only as it affects penetration into the organism (for example, microwaves penetrate better than infrared). Initially, it was believed that low frequency fields that were too weak to cause significant heating could not possibly have any biological effect.[8]

Despite this opinion among researchers, evidence has accumulated that supports the existence of complex biological effects of weaker *non-thermal* electromagnetic fields, (including weak ELF magnetic fields, although the latter does not strictly qualify as EM radiation[8][9][10]), and modulated RF and microwave fields.[11][12][13] Fundamental mechanisms of the interaction between biological material and electromagnetic fields at non-thermal levels are not fully understood.[8]

The World Health Organization has classified radio frequency electromagnetic radiation as Group 2B - possibly carcinogenic.[14][15] This group contains possible carcinogens that have weaker evidence, at the same level as coffee and automobile exhaust. For example, epidemiological studies looking for a relationship between cell phone use and brain cancer development, have been largely inconclusive, save to demonstrate that the effect, if it exists, cannot be a large one.

At higher frequencies (visible and beyond), the effects of individual photons begin to become important, as these now have enough energy individually to directly or indirectly damage biological molecules.[16] All UV frequences have been classed as Group 1 carcinogens by the World Health Organization. Ultraviolet radiation from sun exposure is the primary cause of skin cancer.[17][18]

Thus, at UV frequencies and higher (and probably somewhat also in the visible range),[7] electromagnetic radiation does more damage to biological systems than simple heating predicts. This is most obvious in the "far" (or "extreme") ultraviolet. UV, with X-ray and gamma radiation, are referred to as ionizing radiation due to the ability of photons of this radiation to produce ions and free radicals in materials (including living tissue). Since such radiation can severely damage life at energy levels that produce little heating, it is considered far more dangerous (in terms of damage-produced per unit of energy, or power) than the rest of the electromagnetic spectrum.

2.4.7 Derivation from electromagnetic theory

Main article: Electromagnetic wave equation

Electromagnetic waves were predicted by the classical laws of electricity and magnetism, known as Maxwell's equations. Inspection of Maxwell's equations without sources (charges or currents) results in nontrivial solutions of changing electric and magnetic fields. Beginning with Maxwell's equations in free space:

$$\nabla \cdot \mathbf{E} = 0 \tag{1}$$

$$\nabla \times \mathbf{E} = -\frac{\partial \mathbf{B}}{\partial t} \tag{2}$$

$$\nabla \cdot \mathbf{B} = 0 \tag{3}$$

$$\nabla \times \mathbf{B} = \mu_0 \epsilon_0 \frac{\partial \mathbf{E}}{\partial t} \tag{4}$$

where

> ∇ is a vector differential operator (see Del).

One solution,

$$\mathbf{E} = \mathbf{B} = \mathbf{0},$$

is trivial.

For a more useful solution, we utilize vector identities, which work for any vector, as follows:

$$\nabla \times (\nabla \times \mathbf{A}) = \nabla (\nabla \cdot \mathbf{A}) - \nabla^2 \mathbf{A}$$

The curl of equation (2):

$$\nabla \times (\nabla \times \mathbf{E}) = \nabla \times \left(-\frac{\partial \mathbf{B}}{\partial t} \right) \tag{5}$$

Evaluating the left hand side:

$$\nabla \times (\nabla \times \mathbf{E}) = \nabla (\nabla \cdot \mathbf{E}) - \nabla^2 \mathbf{E} = -\nabla^2 \mathbf{E} \tag{6}$$

simplifying the above by using equation (1).

Evaluating the right hand side:

$$\nabla \times \left(-\frac{\partial \mathbf{B}}{\partial t} \right) = -\frac{\partial}{\partial t} (\nabla \times \mathbf{B}) = -\mu_0 \epsilon_0 \frac{\partial^2 \mathbf{E}}{\partial t^2} \tag{7}$$

Equations (6) and (7) are equal, so this results in a vector-valued differential equation for the electric field, namely

Applying a similar pattern results in similar differential equation for the magnetic field:

These differential equations are equivalent to the wave equation:

$$\nabla^2 f = \frac{1}{c_0^2} \frac{\partial^2 f}{\partial t^2}$$

where

> c_0 is the speed of the wave in free space and
> f describes a displacement

Or more simply:

$$\Box f = 0$$

where $\Box$ is d'Alembertian:

$$\Box = \nabla^2 - \frac{1}{c_0^2} \frac{\partial^2}{\partial t^2} = \frac{\partial^2}{\partial x^2} + \frac{\partial^2}{\partial y^2} + \frac{\partial^2}{\partial z^2} - \frac{1}{c_0^2} \frac{\partial^2}{\partial t^2}$$

In the case of the electric and magnetic fields, the speed is:

$$c_0 = \frac{1}{\sqrt{\mu_0 \epsilon_0}}$$

This is the speed of light in vacuum. Maxwell's equations unified the vacuum permittivity ϵ_0, the vacuum permeability μ_0, and the speed of light itself, c_0. This relationship had been discovered by Wilhelm Eduard Weber and Rudolf Kohlrausch prior to the development of Maxwell's electrodynamics, however Maxwell was the first to produce a field theory consistent with waves traveling at the speed of light.

These are only two equations versus the original four, so more information pertains to these waves hidden within Maxwell's equations. A generic vector wave for the electric field.

$$\mathbf{E} = \mathbf{E}_0 f \left(\hat{\mathbf{k}} \cdot \mathbf{x} - c_0 t \right)$$

Here, $\mathbf{E}_0$ is the constant amplitude, f is any second differentiable function, $\hat{\mathbf{k}}$ is a unit vector in the direction of propagation, and $\mathbf{x}$ is a position vector. $f \left(\hat{\mathbf{k}} \cdot \mathbf{x} - c_0 t \right)$ is a generic solution to the wave equation. In other words,

$$\nabla^2 f \left(\hat{\mathbf{k}} \cdot \mathbf{x} - c_0 t \right) = \frac{1}{c_0{}^2} \frac{\partial^2}{\partial t^2} f \left(\hat{\mathbf{k}} \cdot \mathbf{x} - c_0 t \right),$$

for a generic wave traveling in the $\hat{\mathbf{k}}$ direction.

This form will satisfy the wave equation.

$$\nabla \cdot \mathbf{E} = \hat{\mathbf{k}} \cdot \mathbf{E}_0 f' \left(\hat{\mathbf{k}} \cdot \mathbf{x} - c_0 t \right) = 0$$

$$\mathbf{E} \cdot \hat{\mathbf{k}} = 0$$

The first of Maxwell's equations implies that the electric field is orthogonal to the direction the wave propagates.

$$\nabla \times \mathbf{E} = \hat{\mathbf{k}} \times \mathbf{E}_0 f' \left(\hat{\mathbf{k}} \cdot \mathbf{x} - c_0 t \right) = -\frac{\partial \mathbf{B}}{\partial t}$$

$$\mathbf{B} = \frac{1}{c_0} \hat{\mathbf{k}} \times \mathbf{E}$$

The second of Maxwell's equations yields the magnetic field. The remaining equations will be satisfied by this choice of $\mathbf{E}, \mathbf{B}$.

The electric and magnetic field waves in the far-field travel at the speed of light. They have a special restricted orientation and proportional magnitudes, $E_0 = c_0 B_0$, which can be seen immediately from the Poynting vector. The electric field, magnetic field, and direction of wave propagation are all orthogonal, and the wave propagates in the same direction as $\mathbf{E} \times \mathbf{B}$. Also, $\mathbf{E}$ and $\mathbf{B}$ far-fields in free space, which as wave solutions depend primarily on these two Maxwell equations, are in-phase with each other. This is guaranteed since the generic wave solution is first order in both space and time, and the curl operator on one side of these equations results in first-order spatial derivatives of the wave solution, while the time-derivative on the other side of the equations, which gives the other field, is first order in time, resulting in the same phase shift for both fields in each mathematical operation.

From the viewpoint of an electromagnetic wave traveling forward, the electric field might be oscillating up and down, while the magnetic field oscillates right and left. This picture can be rotated with the electric field oscillating right and left and the magnetic field oscillating down and up. This is a different solution that is traveling in the same direction. This arbitrariness in the orientation with respect to propagation direction is known as polarization. On a quantum level, it is described as photon polarization. The direction of the polarization is defined as the direction of the electric field.

More general forms of the second-order wave equations given above are available, allowing for both non-vacuum propagation media and sources. Many competing derivations exist, all with varying levels of approximation and intended applications. One very general example is a form of the electric field equation,[19] which was factorized into a pair of explicitly directional wave equations, and then efficiently reduced into a single uni-directional wave equation by means of a simple slow-evolution approximation.

2.4.8   See also

- Antenna (radio)
- Antenna measurement
- Bioelectromagnetism
- Bolometer
- Control of electromagnetic radiation
- Electromagnetic field
- Electromagnetic pulse
- Electromagnetic radiation and health
- Electromagnetic spectrum
- Electromagnetic wave equation

- Evanescent wave coupling

- Finite-difference time-domain method

- Helicon

- Impedance of free space

- Light

- Maxwell's equations

- Near and far field

- Radiant energy

- Radiation reaction

- Risks and benefits of sun exposure

- Sinusoidal plane-wave solutions of the electromagnetic wave equation

2.4.9 References

[1] Carmichael, H. J. "Einstein and the Photoelectric Effect" (PDF). Quantum Optics Theory Group, University of Auckland. Retrieved 22 December 2009.

[2] Thorn, J. J.; Neel, M. S.; Donato, V. W.; Bergreen, G. S.; Davies, R. E.; Beck, M. (2004). "Observing the quantum behavior of light in an undergraduate laboratory" (PDF). *American Journal of Physics* **72** (9): 1210. Bibcode:2004AmJPh..72.1210T. doi:10.1119/1.1737397.

[3] Paul M. S. Monk (2004). *Physical Chemistry*. John Wiley and Sons. p. 435. ISBN 978-0-471-49180-4.

[4] Weinberg, S. (1995). *The Quantum Theory of Fields* **1**. Cambridge University Press. pp. 15–17. ISBN 0-521-55001-7.

[5] Philosophical Transactions of the Royal Society of London, Vol. 90 (1800), pp. 284-292, http://www.jstor.org/stable/info/107057

[6] James Jeans (1947) The Growth of Physical Science, link from Internet Archive

[7] Liebel, F.; Kaur, S.; Ruvolo, E.; Kollias, N.; Southall, M. D. (2012). "Irradiation of Skin with Visible Light Induces Reactive Oxygen Species and Matrix-Degrading Enzymes". *Journal of Investigative Dermatology* **132** (7): 1901–1907. doi:10.1038/jid.2011.476. PMID 22318388.

[8] Binhi, Vladimir N; Repiev, A & Edelev, M (translators from Russian) (2002). *Magnetobiology: Underlying Physical Problems*. San Diego: Academic Press. pp. 1–16. ISBN 978-0-12-100071-4. OCLC 49700531.

[9] Delgado, J. M.; Leal, J.; Monteagudo, J. L.; Gracia, M. G. (1982). "Embryological changes induced by weak, extremely low frequency electromagnetic fields". *Journal of anatomy* **134** (Pt 3): 533–551. PMC 1167891. PMID 7107514.

[10] Harland, J. D.; Liburdy, R. P. (1997). "Environmental magnetic fields inhibit the antiproliferative action of tamoxifen and melatonin in a human breast cancer cell line". *Bioelectromagnetics* **18** (8): 555–562. doi:10.1002/(SICI)1521-186X(1997)18:8<555::AID-BEM4>3.0.CO;2-1. PMID 9383244.

[11] Aalto, S.; Haarala, C.; Brück, A.; Sipilä, H.; Hämäläinen, H.; Rinne, J. O. (2006). "Mobile phone affects cerebral blood flow in humans". *Journal of Cerebral Blood Flow & Metabolism* **26** (7): 885–890. doi:10.1038/sj.jcbfm.9600279. PMID 16495939.

[12] Cleary, S. F.; Liu, L. M.; Merchant, R. E. (1990). "In vitro lymphocyte proliferation induced by radio-frequency electromagnetic radiation under isothermal conditions". *Bioelectromagnetics* **11** (1): 47–56. doi:10.1002/bem.2250110107. PMID 2346507.

[13] Ramchandani, P. (2004). "Prevalence of childhood psychiatric disorders may be underestimated". *Evidence-based mental health* **7** (2): 59. doi:10.1136/ebmh.7.2.59. PMID 15107355.

[14] IARC classifies Radiofrequency Electromagnetic Fields as possibly carcinogenic to humans. World Health Organization. 31 May 2011

[15] "Trouble with cell phone radiation standard". *CBS News.*

[16] See PMID 22318388 for evidence of quantum damage from visible light via reactive oxygen species generated in skin. This happens also with UVA. With UVB, the damage to DNA becomes direct, with photochemical formation of pyrimidine dimers.

[17] Narayanan, DL; Saladi, RN; Fox, JL (September 2010). "Ultraviolet radiation and skin cancer". *International Journal of Dermatology* **49** (9): 978–86. doi:10.1111/j.1365-4632.2010.04474.x. PMID 20883261.

[18] Saladi, RN; Persaud, AN (January 2005). "The causes of skin cancer: a comprehensive review". *Drugs of today (Barcelona, Spain : 1998)* **41** (1): 37–53. doi:10.1358/dot.2005.41.1.875777. PMID 15753968.

[19] Kinsler, P. (2010). "Optical pulse propagation with minimal approximations". *Phys. Rev. A* **81**: 013819. arXiv:0810.5689. Bibcode:2010PhRvA..81a3819K. doi:10.1103/PhysRevA.81.013819.

2.4.10 Further reading

- Hecht, Eugene (2001). *Optics* (4th ed.). Pearson Education. ISBN 0-8053-8566-5.

- Serway, Raymond A.; Jewett, John W. (2004). *Physics for Scientists and Engineers* (6th ed.). Brooks Cole. ISBN 0-534-40842-7.

- Tipler, Paul (2004). *Physics for Scientists and Engineers: Electricity, Magnetism, Light, and Elementary Modern Physics* (5th ed.). W. H. Freeman. ISBN 0-7167-0810-8.

- Reitz, John; Milford, Frederick; Christy, Robert (1992). *Foundations of Electromagnetic Theory* (4th ed.). Addison Wesley. ISBN 0-201-52624-7.

- Jackson, John David (1999). *Classical Electrodynamics* (3rd ed.). John Wiley & Sons. ISBN 0-471-30932-X.

- Allen Taflove and Susan C. Hagness (2005). *Computational Electrodynamics: The Finite-Difference Time-Domain Method, 3rd ed.* Artech House Publishers. ISBN 1-58053-832-0.

2.4.11 External links

- Electromagnetism – a chapter from an online textbook

- *Electromagnetic Waves from Maxwell's Equations* on Project PHYSNET.

- Radiation of atoms? e-m wave, Polarisation, ...

- An Introduction to The Wigner Distribution in Geometric Optics

- The windows of the electromagnetic spectrum, on Astronoo

- Introduction to light and electromagnetic radiation course video from the Khan Academy

- Lectures on electromagnetic waves course video and notes from MIT Professor Walter Lewin

- Encyclopedia Britannica Electromagnetic Radiation

- Physics for the 21st Century Early Unification for Electromagnetism Harvard-Smithsonian Center for Astrophysics

2.5 Electromagnetic radiation and health

This article is about the health effects of non-ionizing radiation. For the health effects of ionizing radiation, see radiation poisoning.

Electromagnetic radiation can be classified into two types: ionizing radiation and non-ionizing radiation, based on its capability of ionizing atoms and breaking chemical bonds. Ultraviolet and higher frequencies, such as X-rays or gamma rays are ionizing, and these pose their own special hazards: see *radiation* and *radiation poisoning*. The electricity that comes out of every power socket has associated low-frequency electromagnetic fields.[1] Various kinds of higher-frequency radiowaves are used to transmit information – whether via TV antennas, radio stations or mobile phone base stations.

By far the most common health hazard of radiation is sunburn, which causes over one million new skin cancers annually.[2]

2.5.1 Types of hazards

Electrical hazards

Very strong radiation can induce current capable of delivering an electric shock to persons or animals. It can also overload and destroy electrical equipment. The induction of currents by oscillating magnetic fields is also the way in which solar storms disrupt the operation of electrical and electronic systems, causing damage to and even the explosion of power distribution transformers,[3] blackouts (as occurred in 1989), and interference with electromagnetic signals (*e.g.* radio, TV, and telephone signals).[4]

Fire hazards

Extremely high power electromagnetic radiation can cause electric currents strong enough to create sparks (electrical arcs) when an induced voltage exceeds the breakdown voltage of the surrounding medium (*e.g.* air at 3.0 MV/m). These sparks can then ignite flammable materials or gases, possibly leading to an explosion.

This can be a particular hazard in the vicinity of explosives or pyrotechnics, since an electrical overload might ignite them. This risk is commonly referred to as Hazards of Electromagnetic Radiation to Ordnance (HERO) by the United States Navy (USN). United States Military Standard 464A (MIL-STD-464A) mandates assessment of HERO in a system, but USN document OD 30393 provides design principles and practices for controlling electromagnetic hazards to ordnance.[5]

On the other hand, the risk related to fueling is known as Hazards of Electromagnetic Radiation to Fuel (HERF). NAVSEA OP 3565 Vol. 1 could be used to evaluate HERF, which states a maximum power density of 0.09 W/m² for frequencies under 225 MHz (i.e. 4.2 meters for a 40 W emitter).[5]

Biological hazards

The best understood biological effect of electromagnetic fields is to cause dielectric heating. For example, touching or standing around an antenna while a high-power transmitter is in operation can cause severe burns. These are exactly the kind of burns that would be caused inside a microwave oven.

This heating effect varies with the power and the frequency of the electromagnetic energy. A measure of the heating effect is the specific absorption rate or SAR, which has units of watts per kilogram (W/kg). The IEEE[6] and many national governments have established safety limits for exposure to various frequencies of electromagnetic energy based on SAR, mainly based on ICNIRP Guidelines,[7] which guard against thermal damage.

There are publications which support the existence of complex biological effects of weaker *non-thermal* electromagnetic fields (see Bioelectromagnetics), including weak ELF magnetic fields[8][9] and modulated RF and microwave fields.[10] Fundamental mechanisms of the interaction between biological material and electromagnetic fields at non-thermal levels are not fully understood.[11]

A 2009 study at the University of Basel in Switzerland found that intermittent (but not continuous) exposure of human cells to a 50 Hz electromagnetic field at a flux density of 1 mT (or 10 G) induced a slight but significant increase of DNA fragmentation in the Comet assay.[12] However that level of exposure is already above current established safety exposure limits.

2.5.2 Lighting

Compact fluorescent light bulbs

Compact energy efficient fluorescent light bulbs may emit dangerous levels of Ultraviolet radiation when the protective coating around the phosphor, which creates light inside the bulb, is cracked by mishandling or faulty manufacturing. This cracking of bulb's shielding allows UV rays to escape at levels that could cause burns or even skin cancer. The light generated inside the bulb of a fluorescent light is invisible UV which is converted into visible light by the phosphor coating.[13][14]

LED lights

See also: High CRI LED lighting

Blue light, emitting at wavelengths of 400–500 nanometers, suppresses the production of melatonin produced by the pineal gland. The effect is disruption of a human being's biological clock resulting in poor sleeping and rest periods.[15]

2.5.3 EMR effects on the human body by frequency

Warning sign next to a transmitter with high field strengths

While the most acute exposures to harmful levels of electromagnetic radiation are immediately realized as burns, the health effects due to chronic or occupational exposure may not manifest effects for months or years.

Extremely-low-frequency RF

High-power extremely-low-frequency RF with electric field levels in the low kV/m range are known to induce perceivable currents within the human body that create an annoying tingling sensation. These currents will typically flow to ground through a body contact surface such as the feet, or arc to ground where the body is well insulated.[16][17]

Shortwave frequency RF

Shortwave diathermy heating of human tissue only heats tissues that are good electrical conductors, such as blood vessels and muscle. Adipose tissue (fat) receives little heating by induction fields because an electrical current is not actually going through the tissues.[18]

Radio frequency fields

See also: Mobile phone radiation and health

Apart from some suspicion that the electromagnetic fields emitted by mobile phones may be responsible for an increased risk of glioma and acoustic neuroma, the fields otherwise pose no risk to human health.[19][20] This designation of mobile phone signals as "possibly carcinogenic" by the World Health Organization has often been misinterpreted as indicating that of some measure of risk has been observed – however the designation indicates only that the possibility could not be conclusively ruled out using the available data.[21]

Microwaves

Microwave exposure at low-power levels below the specific absorption rate set by government regulatory bodies is considered harmless non-ionizing radiation and has no effect on the human body. Levels above the specific absorption rate set by the US Federal Communications Commission (FCC) are those they considered to be potentially harmful. ANSI standards for safe exposure levels to RF and microwave radiation are set to a SAR level of 4 W/kg, the threshold before hazardous thermical effects occur due to energy absorption in the body. A safety factor of ten was then incorporated to arrive at the final recommended protection guidelines of a SAR exposure threshold of 0.4 W/kg for RF and microwave radiation. There is disagreement over exactly what levels of RF radiation are safe, particularly with regard to low levels of exposure. Russia and eastern European countries set SAR thresholds for microwaves and RF much lower than western countries.

Two areas of the body, the eyes and the testes, can be particularly susceptible to heating by RF energy because of the relative lack of available blood flow to dissipate the excessive heat load. Laboratory experiments have shown that short-term exposure to high levels of RF radiation (100–200 mW/cm^2) can cause cataracts in rabbits. Temporary sterility, caused by such effects as changes in sperm count and in sperm motility, is possible after exposure of the testes to high-level RF radiation.

Long-term exposure to high-levels of microwaves, is recognized, from experimental animal studies and epidemiological studies in humans, to cause cataracts. The mechanism is unclear but may include changes in heat sensitive enzymes that normally protect cell proteins in the lens. Another mechanism that has been advanced is direct damage to the lens from pressure waves induced in the aqueous humor.

Exposure to sufficiently high-power microwave RF is known to create effects ranging from a burning sensation on the skin and microwave auditory effect, to extreme pain at the mid-range, to physical microwave burns and blistering of skin and internals at high power levels.

Millimeter waves

Recent technology advances in the developments of millimeter wave scanners for airport security and WiGig for Personal area networks have opened the 60 GHz and above microwave band to SAR exposure regulations. Previously, microwave applications in these bands were for point-to-point satellite communication with minimal human exposure. Radiation levels in the millimeter wavelength represent the high microwave band or close to Infrared wavelengths.[22]

Infrared

Infrared wavelengths longer than 750 nm can produce changes in the lens of the eye. Glassblower's cataract is an example of a heat injury that damages the anterior lens capsule among unprotected glass and iron workers. Cataract-like changes can occur in workers who observe glowing masses of glass or iron without protective eyewear for many hours a day.

Another important factor is the distance between the worker and the source of radiation. In the case of arc welding, infrared radiation decreases rapidly as a function of distance, so that farther than three feet away from where welding takes place, it does not pose an ocular hazard anymore but, ultraviolet radiation still does. This is why welders wear tinted glasses and surrounding workers only have to wear clear ones that filter UV.

Visible light

Moderate and high-power lasers are potentially hazardous because they can burn the retina of the eye, or even the skin. To control the risk of injury, various specifications – for example ANSI Z136 in the US, and IEC 60825 internationally – define "classes" of lasers depending on their power and wavelength. These regulations also prescribe required safety measures, such as labeling lasers with specific warnings, and wearing laser safety goggles during operation (see laser safety).

As with its infrared and ultraviolet radiation dangers, welding creates an intense brightness in the visible light spectrum, which may cause temporary flash blindness. Some sources state that there is no minimum safe distance for exposure to these radiation emissions without adequate eye protection.[23]

Ultraviolet

Short-term exposure to strong ultraviolet sunlight causes sunburn within hours of exposure.

Ultraviolet light, specifically UV-B, has been shown to cause cataracts and there is some evidence that sunglasses worn at an early age can slow its development in later life.[24] Most UV light from the sun is filtered out by the atmosphere and consequently airline pilots often have high rates of cataracts because of the increased levels of UV radiation in the upper atmosphere.[25] It is hypothesised that depletion of the ozone layer and a consequent increase in levels of UV light on the ground may increase future rates of cataracts.[26] Note that the lens filters UV light, so once that is removed via surgery, one may be able to see UV light.[27]

Prolonged exposure to ultraviolet radiation from the sun can lead to melanoma and other skin malignancies.[2] Clear evidence establishes ultraviolet radiation, especially the non-ionizing medium wave UVB, as the cause of most non-melanoma skin cancers, which are the most common forms of cancer in the world.[2] UV rays can also cause wrinkles, liver spots, moles, and freckles. In addition to sunlight, other sources include tanning beds, and bright desk lights. Damage is cumulative over one's lifetime, so that permanent effects may not be evident for some time after exposure.[28]

Ultraviolet radiation of wavelengths shorter than 300 nm (actinic rays) can damage the corneal epithelium. This is most commonly the result of exposure to the sun at high altitude, and in areas where shorter wavelengths are readily reflected from bright surfaces, such as snow, water, and sand. UV generated by a welding arc can similarly cause damage to the cornea, known as "arc eye" or welding flash burn, a form of photokeratitis.

2.5.4 See also

- Background radiation
- BioInitiative Report

- Central nervous system effects from radiation exposure during spaceflight
- Cosmic ray
- COSMOS cohort study
- Electromagnetic hypersensitivity
- Electromagnetism
- EMF measurements
- Health threat from cosmic rays
- Light ergonomics
- Magnetobiology
- Microwave auditory effect
- Microwave News
- Mobile phone radiation and health
- Personal RF safety monitors
- Specific absorption rate
- Wireless electronic devices and health

2.5.5 References

[1] What are electromagnetic fields?, World Health Organization

[2] Cleaver JE, Mitchell DL (2000). "15. Ultraviolet Radiation Carcinogenesis". In Bast RC, Kufe DW, Pollock RE; et al. *Holland-Frei Cancer Medicine* (5th ed.). Hamilton, Ontario: B.C. Decker. ISBN 1-55009-113-1. Retrieved 2011-01-31.

[3] "Solar Storms and You: Human Impacts".

[4] Transcript of "Blackout: The Sun-Earth Connection", Part 4: When Solar Plasma Distorts Earth's Magnetic Field

[5] "Acquisition Safety - Radio Frequency Radiation (RFR) Hazards". Naval Safety Center - United States Navy. Retrieved 2014-07-30.

[6] "Standard for Safety Level with Respect to Human Exposure to Radio Frequency Electromagnetic Fields, 3KHz to 300GHz". *IEEE Std* (IEEE). C95.1-2005. October 2005.

[7] International Commission on Non-Ionizing Radiation Protection (April 1998). "Guidelines for limiting exposure to time-varying electric, magnetic, and electromagnetic fields (up to 300 GHz)" (PDF). *Health Physics* **74** (4): 494–522. PMID 9525427.

[8] Delgado JM, Leal J, Monteagudo JL, Gracia MG (May 1982). "Embryological changes induced by weak, extremely low frequency electromagnetic fields". *Journal of Anatomy* **134** (3): 533–51. PMC 1167891. PMID 7107514.

[9] Harland JD, Liburdy RP (1997). "Environmental magnetic fields inhibit the antiproliferative action of tamoxifen and melatonin in a human breast cancer cell line". *Bioelectromagnetics* **18** (8): 555–62. doi:10.1002/(SICI)1521-186X(1997)18:8<555::AID-BEM4>3.0.CO;2-1. PMID 9383244.

[10] Aalto S, Haarala C, Brück A, Sipilä H, Hämäläinen H, Rinne JO (July 2006). "Mobile phone affects cerebral blood flow in humans". *Journal of Cerebral Blood Flow and Metabolism* **26** (7): 885–90. doi:10.1038/sj.jcbfm.9600279. PMID 16495939.

[11] Binhi, Vladimir N; Repiev, A & Edelev, M (translators from Russian) (2002). *Magnetobiology: underlying physical problems*. San Diego: Academic Press. pp. 1–16. ISBN 978-0-12-100071-4. OCLC 49700531.

[12] Focke F, Schuermann D, Kuster N, Schär P (November 2009). "DNA fragmentation in human fibroblasts under extremely low frequency electromagnetic field exposure". *Mutation Research* **683** (1–2): 74–83. doi:10.1016/j.mrfmmm.2009.10.012. PMID 19896957.

[13] Mironava, T.; Hadjiargyrou, M.; Simon, M.; Rafailovich, M. H. (20 Jul 2012). "The Effects of UV Emission from Compact Fluorescent Light Exposure on Human Dermal Fibroblasts and Keratinocytes In Vitro". *Photochemistry and Photobiology* **88**: 1497–1506. doi:10.1111/j.1751-1097.2012.01192.x.

[14] Nicole, Wendee (October 2012). "Ultraviolet Leaks from CFLs". *Environ Health Perspect.* **120** (10): a387. doi:10.1289/ehp.120-a387. PMC 3491932. PMID 23026199.

[15] "Exposure to 'white' light LEDs appears to suppress body's production of melatonin more than certain other lights, research suggests". *ScienceDaily*.

[16] Limits of Human Exposure to Radiofrequency Electromagnetic Fields in the Frequency Range from 3 kHz to 300 GHz, Canada Safety Code 6, page 63

[17] Extremely Low Frequency Fields Environmental Health Criteria Monograph No.238, chapter 5, page 121, WHO

[18] ReBound shortwave diathermy (SWD)

[19] "IARC classifies radiofrequency electromagnetic fields as possibly carcinogenic to humans" (PDF) (Press release). 2011-05-31. Retrieved 2011-08-01.

[20] Baan R, Grosse Y, Lauby-Secretan B, El Ghissassi F, Bouvard V, Benbrahim-Tallaa L, Guha N, Islami F, Galichet L, Straif K, on behalf of the WHO International Agency for Research on Cancer Monograph Working Group (1 July 2011). "Carcinogenicity of radiofrequency electromagnetic fields". *The Lancet Oncology* **12** (7): 624–6. doi:10.1016/S1470-2045(11)70147-4. PMID 21845765. (free, registration required)

[21] Boice JD Jr, Tarone RE (2011). "Cell phones, cancer, and children". *Journal of the National Cancer Institute* **103** (16): 1211–3. doi:10.1093/jnci/djr285. PMID 21795667.

[22] Characterization of 60GHz Millimeter-Wave Focusing Beam for Living-Body Exposure Experiments, Tokoyo University of Technology, Masaki KOUZAI et al., 2009

[23] "What is the minimum safe distance from the welding arc above which there is no risk of eye damage?". The Welding Institute (TWI Global). Retrieved 10 March 2014.

[24] Sliney DH (1994). "UV radiation ocular exposure dosimetry". *Doc Ophthalmol* **88** (3–4): 243–54. doi:10.1007/bf01203678. PMID 7634993.

[25] Rafnsson, V; Olafsdottir E; Hrafnkelsson J; Sasaki H; Arnarsson A; Jonasson F (2005). "Cosmic radiation increases the risk of nuclear cataract in airline pilots: a population-based case-control study". *Arch Ophthalmol* **123** (8): 1102–5. doi:10.1001/archopht.123.8.1102. PMID 16087845.

[26] Dobson, R. (2005). "Ozone depletion will bring big rise in number of cataracts". *BMJ* **331** (7528): 1292–1295. doi:10.1136/bmj.331.7528.1292-d. PMC 1298891.

[27] Komarnitsky. "Case study of ultraviolet vision after IOL removal for Cataract Surgery".

[28] "UV Exposure & Your Health". UV Awareness. Retrieved 10 March 2014.

2.5.6 External links

- Information page on electromagnetic fields at the World Health Organization web site

- Biological Effects of Power Frequency Electric and Magnetic Fields (May 1989) (over 100 pages)

- CDC - Electric and Magnetic Fields - NIOSH Workplace Safety and Health Topic

2.6 International Agency for Research on Cancer

The **International Agency for Research on Cancer** (**IARC**; French: *Centre international de Recherche sur le Cancer, CIRC*) is an intergovernmental agency forming part of the World Health Organization of the United Nations.

Its main offices are in Lyon, France. Its role is to conduct and coordinate research into the causes of cancer. It also collects and publishes surveillance data regarding the occurrence of cancer worldwide.[1] It maintains a series of monographs on the carcinogenic risks to humans posed by a variety of agents, mixtures and exposures.[2] Following its inception, IARC received numerous requests for lists of known and suspected human carcinogens. In 1970, the IARC Advisory Committee recommended that expert groups prepare a compendium on carcinogenic chemicals, and it began publishing its monographs series with this aim in mind.[3]

On October 26, 2015, the International Agency for Research on Cancer of the World Health Organization reported that eating processed meat (eg, bacon, ham, hot dogs, sausages) or red meat was linked to some cancers.[4][5][6]

2.6.1 IARC categories

The IARC categorizes agents, mixtures and exposures into five categories.[7]

- Group 1: carcinogenic to humans.

- Group 2A: probably carcinogenic to humans.

- Group 2B: possibly carcinogenic to humans.

- Group 3: not classifiable as to carcinogenicity in humans.

- Group 4: probably not carcinogenic to humans. Only one substance – caprolactam – has been both assessed for carcinogenicity by the IARC and placed in this category.

2.6.2 Industry influence and transparency

Critics of the IARC have stated that it has become susceptible to industry influence and suffers from a lack of transparency. Lorenzo Tomatis, its director from 1982 to 1993, was "barred from entering the building" in 2003 after "accusing the IARC of softpedaling the risks of industrial chemicals"[8] in a 2002 article.[9] In 2003 thirty public-health scientists signed a letter targeting conflicts of interest and the lack of transparency.[8] The IARC rejected these criticisms, and there was hope that the controversy would "die down" after Paul Kleihues (Director from 1994) retired in 2004 and Peter Boyle became the new director, followed by Christopher Wild since 2009.[8]

Tomatis focused on the IARC monographs which rate chemical's carcinogenicity, and cited several cases in his 2002 critique. In 1998 a panel voted 17-13 to rate 1,3-butadiene a carcinogen. A second vote which Tomatis called "highly irregular" occurred after "industry observers schmoozed with the panelists and one panelist left the meeting", and a 15-14 vote downgraded the chemical to a "possible carcinogen".[8] Joan Denton, director of California's

Office of Environmental Health Hazard Assessment, made accusations in relation to styrene in 2002,[8] and Michael Jacobsen of the Center for Science in the Public Interest criticized the inclusion of industry observers in a saccharin panel, who were allowed to vote.[8] Tomatis has also highlighted DEHP.[10] In defense of the IARC, Kleihues noted that only 17 of 410 of the working-group participants were consultants to industry and these people never served as chairs. He said that "people who receive funds from affected agencies do not vote", and further noted that industry-funded scientists are important because industry often funds studies.[8]

IARC's secrecy led a *Lancet Oncology* editorial to warn of the agency's eroding reputation. As of 2003 the IARC did not release details of disputed votes, did not release the financial disclosure forms required of panelists, or the names of the panelists until the panel is complete. Individuals being considered for the new director are released only to representatives from the 16 member countries. Kleihues and other agency officials defend the IARC, with Kleihues noting that procedures and names are listed on the finished monographs, and said names are not released to avoid political pressures. The IARC was considering new transparency disclosures such as a "narrative" explaining disputed votes.[8]

2.6.3 Criticism

The agency and its "confusing" category system have been criticized for repeatedly misleading the public.[11]

2.6.4 See also

- European Organisation for Research and Treatment of Cancer (EORTC)

- National Cancer Institute (USA)

- Air pollution

- Genotoxic

- Mutagen

- Toxicology

2.6.5 References

[1] http://www-dep.iarc.fr/

[2] http://monographs.iarc.fr/

[3] http://monographs.iarc.fr/ENG/Preamble/CurrentPreamble.pdf

[4] Staff (October 26, 2015). "World Health Organization - IARC Monographs evaluate consumption of red meat and processed meat" (PDF). *International Agency for Research on Cancer*. Retrieved October 26, 2015.

[5] Hauser, Christine (October 26, 2015). "W.H.O. Report Links Some Cancers With Processed or Red Meat". *New York Times*. Retrieved October 26, 2015.

[6] Staff (October 26, 2015). "Processed meats do cause cancer - WHO". *BBC News*. Retrieved October 26, 2015.

[7] "<<Introduction>>". International Programme on Chemical Safety. January 1999. Retrieved 2010-05-16.

[8] Ferber D (July 2003). "Carcinogens. Lashed by critics, WHO's cancer agency begins a new regime". *Science* **301** (5629): 36–7. doi:10.1126/science.301.5629.36. PMID 12843372.

[9] Tomatis L (2002). "The IARC monographs program: changing attitudes towards public health". *Int J Occup Environ Health* **8** (2): 144–52. doi:10.1179/107735202800338993. PMID 12019681.

[10] CBC Markeplace. (2003). Controversy at IARC.

[11] Ed Yong (October 26, 2015). "Why is the World Health Organization so bad at communicating cancer risk?". *The Atlantic*. Retrieved October 26, 2015.

2.6.6 External links

- Official website

Coordinates: 45°44′37″N 4°52′34″E / 45.7435°N 4.8761°E

2.7 List of IARC Group 2B carcinogens

Substances, mixtures and exposure circumstances in this list have been classified by the International Agency for Research on Cancer (IARC) as **Group 2B**: *The agent (mixture) is possibly carcinogenic to humans. The exposure circumstance entails exposures that are possibly carcinogenic to humans.* This category is used for agents, mixtures and exposure circumstances for which there is limited evidence of carcinogenicity in humans and less than sufficient evidence of carcinogenicity in experimental animals. It may also be used when there is inadequate evidence of carcinogenicity in humans but there is sufficient evidence of carcinogenicity in experimental animals. In some instances, an agent, mixture or exposure circumstance for which there is inadequate evidence of carcinogenicity in humans but limited evidence of carcinogenicity in experimental animals together

with supporting evidence from other relevant data may be placed in this group. Further details can be found in the preamble to the IARC Monographs.

2.7.1 Agents and groups of agents

2.7.2 Mixtures

- Bitumens, extracts of steam-refined and air-refined
- Carrageenan, degraded
- Chlorinated paraffins of average carbon chain length C_{12} and average degree of chlorination approximately 60%
- Coffee (urinary bladder[2], large bowel[3])
- Diesel fuel, marine
- Engine exhaust, gasoline
- Fuel oils, residual (heavy)
- Gasoline
- Magenta dyes (CI Basic Red and fuschins)
- Pickled vegetables (traditional in Asia)
- Polybrominated biphenyls [Firemaster BP-6, 59536-65-1]
- Toxaphene (Polychlorinated camphenes)
- Toxins derived from *Fusarium moniliforme*
- Welding fumes

2.7.3 Exposure circumstances

- Carpentry and joinery
- Cobalt metal without tungsten carbide
- Dry cleaning (occupational exposure as)
- Firefighting (occupational exposure as)
- Printing processes (occupational exposure as)
- Talc-based body powders (perineal use of)
- Textile manufacturing industry (work in)

2.7.4 Notes

- ^1 Evaluated as a group.

- ^2 There is limited evidence in humans that coffee drinking is carcinogenic in the urinary bladder.

- ^3 There is some evidence of an inverse relationship between coffee drinking and cancer of the large bowel.

2.7.5 External links

- Agents Classified by the IARC Monographs, International Agency for Research on Cancer

- IARC Monographs - Classifications - by Group

- IARC Monographs on the Evaluation of Carcinogenic Risks to Humans, Volume 51: Coffee, Tea, Mate, Methylxanthines and Methylglyoxal

2.7.6 References

[1] Press release No 208, 31 May 2011, IARC classifies Radiofrequency Electromagnetic Fields as possibly carcinogenic to humans

2.8 Precautionary principle

The **precautionary principle** or precautionary approach to risk management states that if an action or policy has a suspected risk of causing harm to the public or to the environment, in the absence of scientific consensus that the action or policy is not harmful, the burden of proof that it is *not* harmful falls on those taking an action.

The principle is used by policy makers to justify discretionary decisions in situations where there is the possibility of harm from making a certain decision (e.g. taking a particular course of action) when extensive scientific knowledge on the matter is lacking. The principle implies that there is a social responsibility to protect the public from exposure to harm, when scientific investigation has found a plausible risk. These protections can be relaxed only if further scientific findings emerge that provide sound evidence that no harm will result.

In some legal systems, as in the law of the European Union, the application of the precautionary principle has been made a statutory requirement in some areas of law.

Regarding international conduct, the first endorsement of the principle was in 1982 when the World Charter for Nature was adopted by the United Nations General Assembly, while its first international implementation was in 1987

through the Montreal Protocol. Soon after, the principle integrated with many other legally binding international treaties such as the Rio Declaration and Kyoto Protocol.

2.8.1 Origins and theory

The term "precautionary principle" is generally considered to have arisen in English from a translation of the German term *Vorsorgeprinzip* in the 1980s.[1]:31

The concepts underpinning the precautionary principle predate the term's inception. For example, the essence of the principle is captured in a number of cautionary aphorisms such as "an ounce of prevention is worth a pound of cure", "better safe than sorry", and "look before you leap". The precautionary principle may also be interpreted as the evolution of the ancient medical principle of "first, do no harm" to apply to institutions and institutional decision-making processes rather than individuals.

In economics, the precautionary principle has been analysed in terms of the effect on rational decision-making of the interaction of irreversibility and uncertainty. Authors such as Epstein (1980)[2] and Arrow and Fischer (1974)[3] show that irreversibility of possible future consequences creates a quasi-option effect which should induce a "risk-neutral" society to favor current decisions that allow for more flexibility in the future. Gollier et al. (2000)[4] conclude that "more scientific uncertainty as to the distribution of a future risk – that is, a larger variability of beliefs – should induce Society to take stronger prevention measures today."

2.8.2 Formulations

Many definitions of the precautionary principle exist. Precaution may be defined as "*caution in advance*," "*caution practised in the context of uncertainty*," or *informed prudence*. Two ideas lie at the core of the principle:[5]:34

1. an expression of a need by decision-makers to anticipate harm before it occurs. Within this element lies an implicit reversal of the onus of proof: under the precautionary principle it is the responsibility of an activity proponent to establish that the proposed activity will not (or is very unlikely to) result in significant harm.

2. the concept of proportionality of the risk and the cost and feasibility of a proposed action

One of the primary foundations of the precautionary principle, and globally accepted definitions, results from the work of the Rio Conference, or "Earth Summit" in 1992.

Principle #15 of the Rio Declaration notes: "In order to protect the environment, the precautionary approach shall be widely applied by States according to their capabilities. Where there are threats of serious or irreversible damage, lack of full scientific certainty shall not be used as a reason for postponing cost-effective measures to prevent environmental degradation."[6]

The 1998 Wingspread Statement on the Precautionary Principle summarizes the principle this way: "When an activity raises threats of harm to human health or the environment, precautionary measures should be taken even if some cause and effect relationships are not fully established scientifically."[7] The Wingspread Conference on the Precautionary Principle was convened by the Science and Environmental Health Network.[7]

The February 2, 2000 Commission of the European Communities, Communication from the Commission on the Precautionary Principle, noted that, "The precautionary principle is not defined in the Treaty, which prescribes it only once - to protect the environment. But in practice, its scope is much wider, and specifically where preliminary objective scientific evaluation indicates that there are reasonable grounds for concern that the potentially dangerous effects on the environment, human, animal or plant health may be inconsistent with the high level of protection chosen for the Community." [8]:10

The January 29, 2000 Cartagena Protocol on Biosafety says, in regard to controversies over GMOs: "Lack of scientific certainty due to insufficient relevant scientific information . . . shall not prevent the Party of import, in order to avoid or minimize such potential adverse effects, from taking a decision, as appropriate, with regard to the import of the living modified organism in question."[9]:6

2.8.3   Application

The application of the precautionary principle is hampered by both lack of political will, as well as the wide range of interpretations placed on it. One study identified 14 different formulations of the principle in treaties and nontreaty declarations.[10] R.B. Stewart (2002)[11] reduced the precautionary principle to four basic versions:

1. Scientific uncertainty should not automatically preclude regulation of activities that pose a potential risk of significant harm (Non-Preclusion PP).

2. Regulatory controls should incorporate a margin of safety; activities should be limited below the level at which no adverse effect has been observed or predicted (Margin of Safety PP).

3. Activities that present an uncertain potential for significant harm should be subject to best technology available requirements to minimize the risk of harm unless the proponent of the activity shows that they present no appreciable risk of harm (BAT PP).

4. Activities that present an uncertain potential for significant harm should be prohibited unless the proponent of the activity shows that it presents no appreciable risk of harm (Prohibitory PP).

In deciding how to apply the principle, analysis may use a cost-benefit analysis that factors in both the opportunity cost of not acting, and the option value of waiting for further information before acting. One of the difficulties of the application of the principle in modern policy-making is that there is often an irreducible conflict between different interests, so that the debate necessarily involves politics.

Strong vs. weak

Strong precaution holds that regulation is required whenever there is a possible risk to health, safety, or the environment, even if the supporting evidence is speculative and even if the economic costs of regulation are high.[12]:1295–96 In 1982, the United Nations World Charter for Nature gave the first international recognition to the strong version of the principle, suggesting that when "potential adverse effects are not fully understood, the activities should not proceed." The widely publicized Wingspread Declaration, from a meeting of environmentalists in 1998, is another example of the strong version.[13] 'Strong precaution' can also be termed as a "no-regrets" principle, where costs are not considered in preventative action.

Weak precaution holds that lack of scientific evidence does not preclude action if damage would otherwise be serious and irreversible.[14]:1039 Humans practice weak precaution every day, and often incur costs, to avoid hazards that are far from certain: we do not walk in moderately dangerous areas at night, we exercise, we buy smoke detectors, we buckle our seatbelts.[13]

According to a publication by the New Zealand Treasury Department,

> The weak version [of the Precautionary Principle] is the least restrictive and allows preventive measures to be taken in the face of uncertainty, but does not require them (eg, Rio Declaration 1992; United Nations Framework Convention of Climate Change 1992). To satisfy the threshold of harm, there must be some evidence relating to both the likelihood of occurrence and the severity of consequences. Some, but not all, require

consideration of the costs of precautionary measures. Weak formulations do not preclude weighing benefits against the costs. Factors other than scientific uncertainty, including economic considerations, may provide legitimate grounds for postponing action. Under weak formulations, the requirement to justify the need for action (the burden of proof) generally falls on those advocating precautionary action. No mention is made of assignment of liability for environmental harm.

Strong versions justify or require precautionary measures and some also establish liability for environmental harm, which is effectively a strong form of "polluter pays". For example, the Earth Charter states: "When knowledge is limited apply a precautionary approach ... Place the burden of proof on those who argue that a proposed activity will not cause significant harm, and make the responsible parties liable for environmental harm." Reversal of proof requires those proposing an activity to prove that the product, process or technology is sufficiently "safe" before approval is granted. Requiring proof of "no environmental harm" before any action proceeds implies the public is not prepared to accept any environmental risk, no matter what economic or social benefits may arise (Peterson, 2006). At the extreme, such a requirement could involve bans and prohibitions on entire classes of potentially threatening activities or substances (Cooney, 2005). Over time, there has been a gradual transformation of the precautionary principle from what appears in the Rio Declaration to a stronger form that arguably acts as restraint on development in the absence of firm evidence that it will do no harm.[15]

International agreements and declarations

The World Charter for Nature, which was adopted by the UN General Assembly in 1982, was the first international endorsement of the precautionary principle. The principle was implemented in an international treaty as early as the 1987 Montreal Protocol, and among other international treaties and declarations[16] is reflected in the 1992 Rio Declaration on Environment and Development (signed at the United Nations Conference on Environment and Development).

"Principle" vs. "approach"

No introduction to the precautionary princi-

ple would be complete without brief reference to the difference between the precautionary **principle** and the precautionary **approach**. Principle 15 of the Rio Declaration 1992 states that: "in order to protect the environment, the precautionary approach shall be widely applied by States according to their capabilities. Where there are threats of serious or irreversible damage, lack of full scientific certainty shall be not used as a reason for postponing cost-effective measures to prevent environmental degradation." As Garcia (1995) pointed out, "the wording, largely similar to that of the principle, is subtly different in that: (1) it recognizes that there may be differences in local capabilities to apply the approach, and (2) it calls for cost-effectiveness in applying the approach, e.g., taking economic and social costs into account." The 'approach' is generally considered a softening of the 'principle'.

"As Recuerda has noted, the distinction between the ´precautionary principle` and a ´precautionary approach` is diffuse and, in some contexts, controversial. In the negotiations of international declarations, the United States has opposed the use of the term ´principle` because this term has special connotations in legal language, due to the fact that a ´principle of law` is a source of law. This means that it is compulsory, so a court can quash or confirm a decision through the application of the precautionary principle. In this sense, the precautionary principle is not a simple idea or a desideratum but a source of law. This is the legal status of the precautionary principle in the European Union. On the other hand, an ´approach` usually does not have the same meaning, although in some particular cases an approach could be binding. A precautionary approach is a particular ´lens` used to identify risk that every prudent person possesses (Recuerda, 2008)[17]

European Commission On 2 February 2000, the European Commission issued a Communication on the precautionary principle,[8] in which it adopted a procedure for the application of this concept, but without giving a detailed definition of it. Paragraph 2 of article 191 of the Lisbon Treaty states that

"Union policy on the environment shall aim at a high level of protection taking into account the diversity of situations in the various regions of the Union. It shall be based on the precautionary principle and on the principles that preventive action should be taken, that environmental damage

should as a priority be rectified at source and that the polluter should pay."[18]

After the adoption of the European Commission's Communication on the precautionary principle, the principle has come to inform much EU policy, including areas beyond environmental policy. As of 2006 it had been integrated into EU laws "in matters such as general product safety, the use of additives for use in animal nutrition, the incineration of waste, and the regulation of genetically modified organisms."[19]:282–83 Through its application in case law, it has become a "general principle of EU law."[19]:283

USA   On July 18, 2005, the City of San Francisco passed a Precautionary Principle Purchasing ordinance, which requires the city to weigh the environmental and health costs of its $600 million in annual purchases – for everything from cleaning supplies to computers. Members of the Bay Area Working Group on the Precautionary Principle including the Breast Cancer Fund, helped bring this to fruition.

Japan   In 1997, Japan tried to use the consideration of the precautionary principle in a WTO SPS Agreement on the Application of Sanitary and Phytosanitary Measures case, as Japan's requirement to test each variety of agricultural products (apples, cherries, peaches, walnuts, apricots, pears, plums and quinces) for the efficacy of treatment against codling moths was challenged.

This moth is a pest that does not occur in Japan, and whose introduction has the potential to cause serious damage. The United States claimed that it was not necessary to test each variety of a fruit for the efficacy of the treatment, and that this varietal testing requirement was unnecessarily burdensome.

Australia   The most important Australian court case so far, due to its exceptionally detailed consideration of the precautionary principle, is Telstra Corporation Limited v Hornsby Shire Council. The case was heard in the New South Wales Land and Environment Court under Justice CJ Preston (24 April 2006). The Principle was summarised by reference to the NSW *Protection of the Environment Administration Act 1991*, which itself provides a good definition of the principle:

"If there are threats of serious or irreversible environmental damage, lack of full scientific certainty should not be used as a reasoning for postponing measures to prevent environmental degradation. In the application of the principle... decisions should be guided by: (i) careful evaluation to avoid, wherever practicable, serious or irreversible

damage to the environment; and (ii) an assessment of risk-weighted consequence of various options".

The most significant points of Justice Preston's decision are the following findings:

1. The principle and accompanying need to take precautionary measures is "triggered" when two prior conditions exist: a threat of serious or irreversible damage, and scientific uncertainty as to the extent of possible damage.

2. Once both are satisfied, "a proportionate precautionary measure may be taken to avert the anticipated threat of environmental damage, but it should be proportionate."

3. The threat of serious or irreversible damage should invoke consideration of five factors: the scale of threat (local, regional etc.); the perceived value of the threatened environment; whether the possible impacts are manageable; the level of public concern, and whether there is a rational or scientific basis for the concern.

4. The consideration of the level of scientific uncertainty should involve factors which may include: what would constitute sufficient evidence; the level and kind of uncertainty; and the potential to reduce uncertainty.

5. The principle shifts the burden of proof. If the principle applies, the burden shifts: "a decision maker must assume the threat of serious or irreversible environmental damage is... a reality [and] the burden of showing this threat... is negligible reverts to the proponent..."

6. The precautionary principle invokes preventative action: "the principle permits the taking of preventative measures without having to wait until the reality and seriousness of the threat become fully known".

7. "The precautionary principle should not be used to try to avoid all risks."

8. The precautionary measures appropriate will depend on the combined effect of "the degree of seriousness and irreversibility of the threat and the degree of uncertainty... the more significant and uncertain the threat, the greater...the precaution required". "...measures should be adopted... proportionate to the potential threats".

Corporate   The Body Shop International, a UK-based cosmetics company, recently included the Precautionary Principle in their 2006 Chemicals Strategy.

Environment and health

Fields typically concerned by the precautionary principle are the possibility of:

- Global warming or abrupt climate change in general

- Extinction of species

- Introduction of new and potentially harmful products into the environment, threatening biodiversity (e.g., genetically modified organisms)

- Threats to public health, due to new diseases and techniques (e.g., AIDS transmitted through blood transfusion)

- Long-term effects of new technologies (e.g. health concerns regarding radiation from cell phones and other electronics communications devices)

- Persistent or acute pollution (asbestos, endocrine disruptors...)

- Food safety (e.g., Creutzfeldt-Jakob disease)

- Other new biosafety issues (e.g., artificial life, new molecules)

The precautionary principle is often applied to biological fields because changes cannot be easily contained and have the potential of being global. The principle has less relevance to contained fields such as aeronautics, where the few people undergoing risk have given informed consent (e.g., a test pilot). In the case of technological innovation, containment of impact tends to be more difficult if that technology can self-replicate. Bill Joy emphasized the dangers of replicating genetic technology, nanotechnology, and robotic technology in his article in *Wired Magazine*, "Why the future doesn't need us", though he does not specifically cite the precautionary principle. The application of the principle can be seen in the public policy of requiring pharmaceutical companies to carry out clinical trials to show that new medications are safe.

Oxford based philosopher Nick Bostrom discusses the idea of a future powerful superintelligence, and the risks that we/it face should it attempt to gain atomic level control of matter.[20]

Application of the principle modifies the status of innovation and risk assessment: it is not the risk that must be avoided or amended, but a potential risk that must be prevented. Thus, in the case of regulation of scientific research, there is a third party beyond the scientist and the regulator: the consumer.

In an analysis concerning application of the precautionary principle to nanotechnology, Chris Phoenix and Mike Treder posit that there are *two forms* of the principle, which they call the "strict form" and the "active form". The former "requires inaction when action might pose a risk", while the latter means "choosing less risky alternatives when they are available, and [...] taking responsibility for potential risks." Thomas Alured Faunce has argued for stronger application of the precautionary principle by chemical and health technology regulators particularly in relation to TiO_2 and ZnO nanoparticles in sunscreens, biocidal nanosilver in waterways and products whose manufacture, handling or recycling exposes humans to the risk of inhaling multi-walled carbon nanotubes.[21]

Resource management

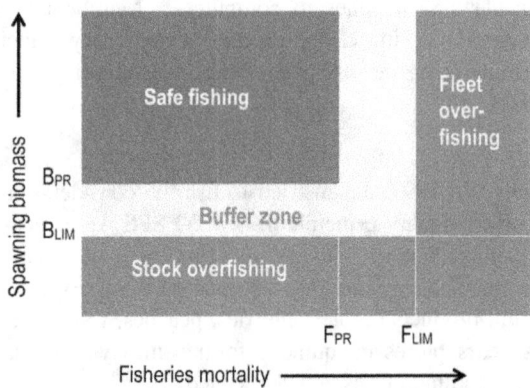

The Traffic Light colour convention, showing the concept of Harvest Control Rule (HCR), specifying when a rebuilding plan is mandatory in terms of precautionary and limit reference points for spawning biomass and fishing mortality rate.

Several natural resources like fish stocks are now managed by precautionary approach, through Harvest Control Rules (HCR) based upon the precautionary principle. The figure indicates how the principle is implemented in the cod fisheries management proposed by the International Council for the Exploration of the Sea.

In classifying endangered species, the precautionary principle means that if there is doubt about an animal's or plant's exact conservation status, the one that would cause the strongest protective measures to be realized should be chosen. Thus, a species like the silvery pigeon that might exist in considerable numbers and simply be under-recorded or might just as probably be long extinct is not classified as "data deficient" or "extinct" (which both do not require any protective action to be taken), but as "critically endangered" (the conservation status that confers the need for the strongest protection), whereas the increasingly rare, but probably not yet endangered emerald starling is classified as "data deficient", because there is urgent need for research to clarify its status rather than for conservation action to save it from extinction.

If, for example, a large ground-water body that many people use for drinking water is contaminated by bacteria (e-coli 0157 H7, campylobacter or leptospirosis) and the source of contamination is strongly suspected to be dairy cows but the exact science is not yet able to provide absolute proof, the cows should be removed from the environment until they are proved, by the dairy industry, not to be the source or until that industry ensures that such contamination will not recur.

2.8.4 Criticisms

Critics of the principle use arguments similar to those against other formulations of technological conservatism.

Internal inconsistency - applying strong PP risks causing harm

Strong formulations of the precautionary principle, without regard to its most basic provisions that it is to be applied only where risks are potentially high AND not easily calculable, applied to the principle itself as a policy decision, may rule out its own use.[14]:26ff The reason suggested is that preventing innovation from coming to market means that only current technology may be used, and current technology itself may cause harm or leave needs unmet; there is a risk of causing harm by blocking innovation.[22][23] As Michael Crichton wrote in his novel, *State of Fear*: "The 'precautionary principle', properly applied, forbids the precautionary principle."[24] For example, forbidding nuclear power plants based on concerns about risk means continuing to rely on power plants that burn fossil fuels, which continue to release greenhouse gases.[14]:27 In another example, the Hazardous Air Pollutant provisions in the 1990 amendments to the U.S. Clean Air Act are an example of the Precautionary Principle where the onus is now on showing a listed compound is harmless. Under this rule no distinction is made between those air Pollutants that provide a higher or lower risk, so operators tend to choose less-examined agents that are not on the existing list.[25]

Blocking innovation and progress generally

Because applications of strong formulations of the PP can be used to block innovation, a technology which brings advantages may be banned by PP because of its potential for negative impacts, leaving the positive benefits unrealized.[26][27]:201

The precautionary principle has been ethically questioned on the basis that its application could block progress in developing countries.[28]

Vagueness and plausibility

The PP calls for inaction in the face of scientific uncertainty, but some formulations do not specify the minimal threshold of plausibility of risk that acts as a "triggering" condition, so that any indication that a proposed product or activity might harm health or the environment is sufficient to invoke the principle.[29][30] In *Sancho vs. DOE*, Helen Gillmor, Senior District Judge, wrote in a dismissal of Wagner's lawsuit which included a popular[31] worry that the LHC could cause "destruction of the earth" by a black hole:

> Injury in fact requires some "credible threat of harm." *Cent. Delta Water Agency v. United States*, 306 F.3d 938, 950 (9th Cir. 2002). At most, Wagner has alleged that experiments at the Large Hadron Collider (the "Collider") have "potential adverse consequences." Speculative fear of future harm does not constitute an injury in fact sufficient to confer standing. *Mayfield*, 599 F.3d at 970.[32]

2.8.5 See also

2.8.6 References

[1] Sonja Boehmer Christiansen. The Precautionary Principle in Germany: Enabling Government. Chapter 2 in Interpreting the Precautionary Principle, eds Tim O'Riordan and James Cameron Earthscan Publications Ltd, 1994

[2] Epstein, L.S. (1980). "Decision-making and the temporal resolution of uncertainty". *International Economic Review* **21** (2): 269–283. doi:10.2307/2526180. JSTOR 2526180.

[3] Arrow, K.J. and Fischer, A.C. (1974). "Environmental preservation, uncertainty and irreversibility". *Quarterly Journal of Economics* **88** (2): 312–9. doi:10.2307/1883074. JSTOR 1883074.

[4] Gollier, Christian, Bruno Jullien & Nicolas Treich (2000). "Scientific Progress and Irreversibility: An Economic Interpretation of the 'Precautionary Principle'". *Journal of Public Economics* **75** (2): 229–253. doi:10.1016/S0047-2727(99)00052-3.

[5] Andrew Jordan & Timothy O'Riordan. Chapter 3, The precautionary principle: a legal and policy history, in The precautionary principle: protecting public health, the environment and the future of our children. Edited by: Marco Martuzzi and Joel A. Tickner. World Health Organization 2004

[6] UNEP. "Rio Declaration on Environment and Development". Retrieved 29 October 2014.

[7] Staff, Science and Environmental Health Network. January 26, 1998 Wingspread Conference on the Precautionary Principle

[8] Commission of the European Communities. February 2, 2000 Communication From The Commission on the Precautionary Principle

[9] Official text of the Cartagena Protocol

[10] Foster, Kenneth R.; Vecchia, Paolo; Repacholi, Michael H. (2000). "Science and the Precautionary Principle". *Science* **288** (5468): 979–981. doi:10.1126/science.288.5468.979. ISSN 0036-8075. PMID 10841718

[11] Stewart, R.B. (2002). "Environmental Regulatory Decision Making Under Uncertainty". *Research in Law and Economics* **20**: 76.

[12] Sachs, Noah M. (2011). "Rescuing the Strong Precautionary Principle from its Critics" (PDF). *University of Illinois Law Review* **2011** (4): 1285–1338.

[13] "The paralyzing principle: Does the precautionary principle point us in any helpful direction?" Goliath Business Knowledge on Demand. December 2002. http://goliath.ecnext.com/coms2/gi_0199-2593495/The-paralyzing-principle-does-the.html

[14] Mandel, Gregory N.; Gathii, James Thuo (2006). "Cost Benefit Analysis Versus the Precautionary Principle: Beyond Cass Sunstein's Laws of Fear" (PDF). *University of Illinois Law Review* **2006** (5): 1037–1079

[15] "Precautionary Principle: Origins, definitions, and interpretations." Treasury Publication, Government of New Zealand. 2006. http://www.treasury.govt.nz/publications/research-policy/ppp/2006/06-06/05.htm

[16]

[17] Recuerda, M. A. (2008). "Dangerous interpretations of the precautionary principle and the foundational values of the European Union Food Law: Risk versus Risk". *Journal of Food Law & Policy* **4** (1).

[18] Consolidated Version of the Treaty on the Functioning of the European Union article 191, paragraph 2

[19] Recuerda, Miguel A. (2006). "Risk and Reason in the European Union Law". *European Food and Feed Law Review* **5**.

[20] Nick Bostrom 2003 Ethical Issues in Advanced Artificial Intelligence – section 2

[21] Faunce TA, et al. (2008). "Sunscreen Safety: The Precautionary Principle, The Australian Therapeutic Goods Administration and Nanoparticles in Sunscreens" (PDF). *Nanoethics* **2** (3): 231–240. doi:10.1007/s11569-008-0041-z. Archived from the original (PDF) on 2011-02-28.

[22] Brown, Tracey (9 July 2013)The precautionary principle is a blunt instrument *The Guardian*, Retrieved 9 August 2013

[23] Sherry Seethaler. Lies, Damned Lies, and Science: How to Sort through the Noise around Global Warming, the Latest Health Claims, and Other Scientific Controversies FT Press, 2009

[24] Merchant, G et al. Impact of the Precautionary Principle on Feeding Current and Future Generations CAST Issue Paper 52, June 2013

[25] Goldstein BD, Carruth RS (2004). "Implications of the Precautionary Principle: is it a threat to science?". *Int J Occup Med Environ Health* **17** (1): 153–61. PMID 15212219.

[26] Sunstein, Cass R. The Paralyzing Principle: Does the Precautionary Principle Point us in any Helpful Direction? Regulation, Winter 2002–2003, The Cato Institute.

[27] David Deutsch, The Beginning of Infinity Penguin Books (UK), Viking Press (US), 2011. ISBN 978-0-7139-9274-8

[28] Jimenez-Arias, Luis G. (2008). *Biothics and the Environment*. Libros en Red. p. 73.

[29] van den Belt H (July 2003). "Debating the Precautionary Principle: "Guilty until Proven Innocent" or "Innocent until Proven Guilty"?". *Plant Physiol.* **132** (3): 1122–6. doi:10.1104/pp.103.023531. PMC 526264. PMID 12857792.

[30] Bailey, Ronald. Precautionary Tale. Reason. April 1999

[31] Roger Highfield, Science Editor (5 September 2008). "Scientists get death threats over Large Hadron Collider". *Telegraph.co.uk*. Retrieved 29 October 2014.

[32] "LHC lawsuit dismissed by US court". *symmetry magazine*. Retrieved 29 October 2014.

2.8.7 Other publications

- Kai Purnhagen, "The Behavioural Law and Economics of the Precautionary Principle in the EU and its Impact on Internal Market Regulation", Wageningen Working Papers in Law and Governance 2013/04,

- Arrow, K.J.; et al. (1996). "Is There a Role for Cost-Benefit Analysis in Environmental, Health, and Safety Regulation?". *Science* **272** (5259): 221–2. doi:10.1126/science.272.5259.221. PMID 8602504.

- Andorno, Roberto (2004). "The Precautionary Principle: A New Legal Standard for a Technological Age" (PDF). *Journal of International Biotechnology Law* **1**: 11–19. doi:10.1515/jibl.2004.1.1.11.

- Communication from the European Commission on the precautionary principle Brusells (2000)

- European Union (2002), European Union consolidated versions of the treaty on European Union and of the treaty establishing the European community, Official Journal of the European Union, C325, 24 December 2002, Title XIX, article 174, paragraph 2 and 3.

- Greenpeace, "Safe trade in the 21st Century, Greenpeace comprehensive proposals and recommendations for the 4th Ministerial Conference of the World Trade Organisation" pp. 8–9

- Harremoës, Poul, David Gee, Malcolm MacGarvin, Andy Stirling, Jane Keys, Brian Wynne, Sofia Guedes Vaz (October 2002). "The Precautionary Principle in the 20th Century: Late Lessons from Early Warnings — Earthscan, 2002. Review". *Nature* **419** (6906): 433. doi:10.1038/419433a.

- O'Riordan, T. and Cameron, J. (1995), *Interpreting the Precautionary Principle*, London: Earthscan Publications

- Raffensperger, C., and Tickner, J. (eds.) (1999) Protecting Public Health and the Environment: Implementing the Precautionary Principle. Island Press, Washington, DC.

- Recuerda Girela, M.A., (2006), Seguridad Alimentaria y Nuevos Alimentos, Régimen jurídico-administrativo. Thomson-Aranzadi, Cizur Menor.

- Recuerda Girela, M.A., (2006), "Risk and Reason in the European Union Law", European Food and Feed Law Review, 5.

- Ricci PF, Rice D, Ziagos J, Cox LA (April 2003). "Precaution, uncertainty and causation in environmental decisions". *Environ Int* **29** (1): 1–19. doi:10.1016/S0160-4120(02)00191-5. PMID 12605931.

- Sandin, P. "Better Safe than Sorry: Applying Philosophical Methods to the Debate on Risk and the Precautionary Principle," (2004).

- Stewart, R.B. "Environmental Regulatory Decision making under Uncertainty". In An Introduction to the Law and Economics of Environmental Policy: Issues in Institutional Design, Volume 20: 71–126 (2002).

- Sunstein, Cass R. (2005), *Laws of Fear: Beyond the Precautionary Principle*. New York: Cambridge University Press

2.8.8   External links

- A Small Dose of Toxicology

- Bay Area Working Group on the Precautionary Principle

- Roberto Andorno, "The Precautionary Principle: A New Legal Standard for a Technological Age", *Journal of International Biotechnology Law,* 2004, 1, p. 11–19

- Report by the UK Interdepartmental Liaison Group on Risk Assessment, 2002. "The Precautionary Principle: Policy and Application"

- David Appell, *Scientific American*, January 2001: "The New Uncertainty Principle"

- *The Times*, July 27, 2007, Only a reckless mind could believe in safety first

- *The Times*, January 15, 2005, "What is . . . the Precautionary Principle?"

- Bill Durodié, *Spiked*, March 16, 2004: The precautionary principle assumes that prevention is better than cure

- European Environment Agency (2001), Late lessons from early warnings: the precautionary principle 1896–2000

- Applying the Precautionary Principle to Nanotechnology, *Center for Responsible Nanotechnology* 2004

- 1998 *Wingspread Statement on the Precautionary Principle*

- Science and Environmental Health Network, The Precautionary Principle in Action – a Handbook

- Gary E. Marchant, Kenneth L. Mossman: *Arbitrary and Capricious: The Precautionary Principle in the European Union Courts.* American Enterprise Institute Press 2004, ISBN 0-8447-4189-2; free online PDF

- Umberto Izzo, La precauzione nella responsabilità civile. Analisi di un concetto sul tema del danno da contagio per via trasfusionale (e-book reprint) [The Idea of Precaution in Tort Law. Analysis of a Concept against the Backdrop of the Tainted- Blood Litigation], UNITN e-prints, 2007, first edition Padua, Cedam 2004.free online PDF

- The Precautionary Principle Project: Sustainable Development, Natural Resource Management and Biodiversity Conservation

- Better Safe than Sorry: Applying Philosophical Methods to the Debate on Risk and the Precautionary Principle

- Communication from the European Commission on the precautionary principle

- UK Interdepartmental Liaison Group on Risk Assessment (ILGRA): The Precautionary Principle: Policy and Application

- Report of UNESCO's group of experts on the Precautionary Principle (2005)

- Max More (2010), The Perils Of Precaution

2.9 Scientific Committee on Emerging and Newly Identified Health Risks

The **Scientific Committee on Emerging and Newly Identified Health Risks** (**SCENIHR**) is one of the independent scientific committees managed by the Directorate-General for Health and Consumer Protection of the European Commission, which provide scientific advice to the Commission on issues related to consumer products.

2.9.1 Activities

The SCENIHR provides scientific opinions on questions concerning emerging or newly identified risks on non-food products, as well as on broad, complex or multidisciplinary issues requiring a comprehensive assessment of risks to consumer safety or public health not covered by other risk assessment bodies.

Examples of areas of activity include new technologies (such as nanotechnologies), medical devices, antimicrobial resistance, physical risks (such as noise and electromagnetic fields), and methodologies of risk assessment.

2.9.2 Procedures

SCENIHR's scientific advisory procedures are based on the principles of scientific excellence, independence and transparency.

2.9.3 See also

The Directorate-General for Health and Consumer Protection also manages two other independent Scientific Committees on non-food products:

- The Scientific Committee on Consumer Safety (SCCS)

- The Scientific Committee on Health and Environmental Risks (SCHER)

For questions concerning the safety of food products, the European Commission consults the European Food Safety Authority (EFSA).

2.9.4 External links

- Scientific Committee on Emerging and Newly Identified Health Risks

- The Scientific Committees of the Directorate-General for Health and Consumer Protection

- Commission Decision 2008/721/EC of 5 August 2008 setting up an advisory structure of Scientific Committees and experts in the field of consumer safety, public health and the environment (as amended) from EUR-Lex

2.10 Wireless electronic devices and health

For mobile phones, see Mobile phone radiation and health.

The World Health Organization (WHO) has acknowledged the "anxiety and speculation" regarding electromagnetic fields (EMFs) and their alleged effects on public health.[1]

In response to public concern, the WHO established the *International EMF Project* in 1996 to assess the scientific evidence of possible health effects of EMF in the frequency range from 0 to 300 GHz. They have stated that although extensive research has been conducted into possible health effects of exposure to many parts of the frequency spectrum, all reviews conducted so far have indicated that, as long as exposures are below the limits recommended in the ICNIRP (1998) EMF guidelines, which cover the full frequency range from 0–300 GHz, such exposures do not produce any known adverse health effect.[2] Of course, by the very definition of such limits, stronger or more frequent exposures to EMF can be unhealthy, and in fact serve as the basis for electromagnetic weaponry.

International guidelines on exposure levels to microwave frequency EMFs such as ICNIRP limit the power levels of wireless devices and it is uncommon for wireless devices to exceed the guidelines. These guidelines only take into account thermal effects, as nonthermal effects have not been conclusively demonstrated.[3] The official stance of the British Health Protection Agency is that "[T]here is no consistent evidence to date that WiFi and WLANs adversely affect the health of the general population", but also that "...it is a sensible precautionary approach...to keep the situation under ongoing review...".[4]

In 2011, International Agency for Research on Cancer (IARC), an agency of the World Health Organization, classified wireless radiation as Group 2B – possibly carcinogenic. That means that there "could be some risk" of carcinogenicity, so additional research into the long-term, heavy use of wireless devices needs to be conducted.[5]

2.10.1 Exposure difference to mobile phones

Main article: Mobile phone radiation and health

Users of wireless devices are typically exposed for much longer periods than for mobile phones and the strength of wireless devices is not significantly less. Whereas a mobile phone can range from 21 dBm (125 mW) for Power Class 4 to 33 dBm (2W) for Power class 1, a wireless router can range from a typical 15 dBm (30 mW) strength to 27 dBm (500 mW) on the high end.[6]

However, wireless routers are typically located significantly farther away from users' heads than a mobile phone the user is handling, resulting in far less exposure overall. The Health Protection Agency (HPA) claims that if a person spends one year in a location with a Wi-Fi hotspot, they will receive the same dose of radio waves as if they had made a 20-minute call on a mobile phone.[7]

The HPA also acknowledges that due to the mobile phone's adaptive power ability, a DECT cordless phone's radiation could actually exceed the radiation of a mobile phone. The HPA explains that while the DECT cordless phone's radiation has an average output power of 10 mW, it is actually in the form of 100 bursts per second of 250 mW, a strength comparable to some mobile phones.[8]

2.10.2 Wireless LAN

Most wireless LAN equipment is designed to work within predefined standards. Wireless access points are also often close to humans, but the drop off in power over distance is fast, following the inverse-square law.[9] However, wireless laptops are typically used close to humans. WiFi has been anecdotally linked to electromagnetic hypersensitivity, e.g., in Toronto, Canada schoolchildren as well as staff workers of France National Library.[10]

The HPA's position is that "...radio frequency (RF) expo-

sures from WiFi are likely to be lower than those from mobile phones." It also saw "...no reason why schools and others should not use WiFi equipment."[4] In October 2007, the HPA launched a new "systematic" study into the effects of WiFi networks on behalf of the UK government, in order to calm fears that had appeared in the media in a recent period up to that time".[11] Dr Michael Clark, of the HPA, says published research on mobile phones and masts does not add up to an indictment of WiFi.[12][13]

2.10.3 Other devices

Radio frequency in the microwave and radio spectrum is used in a number of practical devices for professional and home use, such as:

- DECT and other cordless phones operating at a wide range of frequencies

- Remote control devices for opening gates, etc.

- Portable two-way radio communication devices, such as walkie-talkies, etc.

- Wireless security (alarm) systems

- Wireless security video cameras

- Radio links between buildings for data communication

- Baby monitors

- Smart meters for electric energy

- Bluetooth and other personal area network devices - e.g., wireless headphones, smart watches.

- Implantable medical devices - pacemakers, implanted defibrillators

In addition, electrical and electronic devices of all kinds emit EM fields around their working circuits, generated by oscillating currents. Humans are in daily contact with computers, video display monitors, TV screens, microwave ovens, fluorescent lamps, electric motors of several kinds (such as washing machines, kitchen appliances [like electric can openers, blenders, and mixers], water pumps, etc.) and many others. A study of bedroom exposure in 2009 showed the highest ELF-EF from bedside lights and the highest ELF-MF from transformer devices, while the highest RF-ELF came from DECT cordless phones and outside cellphone base stations; all exposures were well below International Commission on Non-Ionizing Radiation Protection (ICNIRP) guideline levels.[14]

The highest typical daily exposure, according to a study of 2009, came from cellphone base stations, cellphones and

DECT cordless phones, with the highest exposure locations in trains, airports and buses.[15] The typical background power of electromagnetic fields in the home can vary from zero to 5 milliwatts per meter squared. Long-time effects of these electromagnetic fields on human and animal health are still unknown.

Some implanted medical devices use radio frequency communication - both to report status, and to allow changing device behavior. Emissions from wireless electronic devices can interfere with the functioning of these devices, thereby adversely affecting the health of the user. Users of such implanted devices are usually cautioned to avoid close exposure to other wireless devices.

2.10.4   See also

- Electromagnetic radiation and health

2.10.5   References

[1] "Electromagnetic fields (EMF)". World Health Organization. Retrieved 2008-01-22. "Electromagnetic fields of all frequencies represent one of the most common and fastest growing environmental influences, about which anxiety and speculation are spreading. All populations are now exposed to varying degrees of EMF, and the levels will continue to increase as technology advances."

[2] "WHO EMF Research". World Health Organisation. Retrieved 2012-03-27.

[3] Levitt, B. Blake (1995). *Electromagnetic Fields : a consumer's guide to the issues and how to protect ourselves.* San Diego: Harcourt Brace. pp. 29–38. ISBN 978-0-15-628100-3. OCLC 32199261.

[4] "WiFi Summary". Health Protection Agency. Retrieved 2010-01-09.

[5] "IARC classifies radiofrequency electromagnetic fields as possibly carcinogenic to humans" (PDF). *World Health Organization press release N° 208* (Press release). International Agency for Research on Cancer. 2011-05-31. Retrieved 2011-06-02.

[6] dBm

[7] "Wi-fi health fears are 'unproven'". *BBC News* (BBC). 2007-05-21. Retrieved 2008-01-22.

[8] http://www.hpa.org.uk/Topics/Radiation/ UnderstandingRadiation/InformationSheets/info_ CordlessTelephones/

[9] Foster, Kenneth R (March 2007). "Radiofrequency exposure from wireless LANs utilizing Wi-Fi technology". *Health Physics* **92** (3): 280–289. doi:10.1097/01.HP.0000248117.74843.34. PMID 17293700.

[10] "Ont. parents suspect Wi-Fi making kids sick". *CBC News.* 2010-08-16.

[11] "Health Protection Agency announces further research into use of WiFi". Health Protection Agency. Retrieved 2008-08-28.

[12] Nicki Daniels (11 December 2006). "Wi-fi: should we be worried?". The Times. Retrieved 26 May 2015.

[13] "Bioinitiative Report". Retrieved 5 October 2013.

[14] Tomitsch, Johannes; Dechant, Engelbert; Frank, Wilhelm (2009-09-24). "Survey of electromagnetic field exposure in bedrooms of residences in lower Austria". *Bioelectromagnetics* **31**: 200–8. doi:10.1002/bem.20548. PMID 19780092.

[15] Frei, Patrizia; Mohler, Evelyn; Neubauer, Georg; Theis, Gaston; Bürgi, Alfred; Fröhlich, Jürg; Braun-Fahrländer, Charlotte; Bolte, John; Egger, Matthias; Röösli, Martin (August 2009). "Temporal and spatial variability of personal exposure to radio frequency electromagnetic fields". *Environmental Research* **109** (6): 779–785. doi:10.1016/j.envres.2009.04.015. PMID 19476932.

2.10.6   External links

- WiFi information at the UK Health Protection Agency

Chapter 3

Text and image sources, contributors, and licenses

3.1 Text

- **Mobile phone** *Source:* https://en.wikipedia.org/wiki/Mobile_phone?oldid=687354139 *Contributors:* WojPob, Bryan Derksen, Robert Merkel, Zundark, The Anome, Koyaanis Qatsi, Malcolm Farmer, Ed Poor, Xaonon, Youssefsan, RobertBrook, Hajhouse, Nate Silva, Little guru, William Avery, LionKimbro, Heron, Rsabbatini, Montrealais, Youandme, Tzartzam, Olivier, Edward, Patrick, Infrogmation, Michael Hardy, Tim Starling, Paul Barlow, DopefishJustin, Jtdirl, Pnm, GUllman, Liftarn, Gabbe, Ixfd64, Dcljr, Sannse, Cameron Dewe, 6birc, Tzaquiel, Delirium, Dori, Minesweeper, Goatasaur, Tregoweth, NuclearWinner, Ellywa, Ahoerstemeier, Mac, Nanshu, Jpatokal, Snoyes, Angela, Jdforrester, Kingturtle, Pmolinero~enwiki, RadRafe, Stefan-S, Rossami, Benjaminong, Kwekubo, Andres, Tristanb, Jimregan, Wael Ellithy, GRAHAMUK, Ehn, Arteitle, Reinhard Kraasch, Stephenw32768, Alakon, Chatool, Dysprosia, Fuzheado, WhisperToMe, Selket, Steinsky, Radiojon, Roadmr, DJ Clayworth, Birkett, Tpbradbury, Kierant, Maximus Rex, Furrykef, Pacific1982, Saltine, Nv8200pa, ZeWrestler, Phoebe, Wernher, דוד, Topbanana, Fvw, Stormie, Pakaran, Guppy, Francs2000, Cluth, Owen, Lumos3, Shantavira, RadicalBender, Northgrove, Riddley, Bearcat, Robbot, Paranoid, Dale Arnett, Hankwang, Craig Stuntz, PBS, Jredmond, Gak, RedWolf, Donreed, Moncrief, Psychonaut, TimothyPilgrim, Pelle, Babbage, JustinHall, P0lyglut, Denots, Merovingian, Alexblainelayder, Academic Challenger, Alexaq~enwiki, Rhombus, Jondel, Rasmus Faber, Leedar, Hadal, Cyrius, Alan Liefting, David Gerard, Matthew Stannard, Giftlite, Dbenbenn, Anym, Lcgarcia, DocWatson42, Laudaka, Dinomite, Geeoharee, Lupin, MathKnight, Orangemike, Obli, Monedula, Peruvianllama, Everyking, Capitalistroadster, Curps, Alison, Cantus, Filceolaire, Bsoft, Sundar, Zoney, AlistairMcMillan, Thomas Ludwig, Solipsist, Foobar, Iceberg3k, Khalid hassani, Uzume, Bobblewik, Tagishsimon, Wmahan, Somebody512, Wikiwiki~enwiki, Mackeriv, Utcursch, 159753, SoWhy, Spizzer2, Sohailstyle, CryptoDerk, LiDaobing, Technogeek, Yardcock, Antandrus, Beland, Joeblakesley, OverlordQ, Scottperry, Kusunose, ShakataGaNai, Arsene, CaribDigita, Thevirus, Heman, Rdsmith4, Kesac, OwenBlacker, Maximaximax, Mysidia, Elroch, Mozzerati, Sam Hocevar, Histrion, Jklamo, Gscshoyru, Tooki, Jamespole, Creidieki, Neutrality, Joyous!, Jcw69, Ianneub, Shadowlink1014, Cab88, Kevin Rector, M1ss1ontomars2k4, Damieng, Trevor MacInnis, Acsenray, Moxfyre, Randwicked, Grunt, Canterbury Tail, Andylkl, Qu1che, Bluemask, Zaf, Forschung, Joseph Philipsson, Corti, Mike Rosoft, Myfanwy, Imroy, Discospinster, Rich Farmbrough, KillerChihuahua, Rhobite, Guanabot, Oliver Ruehl, NrDg, Hydrox, DrMac, Selphie, Wrp103, Vsmith, StephanKetz, YUL89YYZ, IlyaHaykinson, Xezbeth, Mjpieters, Saintswithin, Ibagli, Zazou, Mani1, Pavel Vozenilek, Xeper, Paul August, Pnevares, Plumpy, Bender235, ESkog, TerraFrost, Ground, JoeSmack, Janaagaard, Petersam, Violetriga, Evice, Billion, Stebbiv, Yvolution, Brian0918, Tompw, Sockatume, CanisRufus, Torindkflt, FirstPrinciples, MBisanz, El C, Lycurgus, Hayabusa future, Mwanner, Jantangring, PhilHibbs, Mavhc, Sietse Snel, Art LaPella, RoyBoy, Cacophony, G worroll, Mairi, Coolcaesar, 2005, Jpgordon, Alxndr, Bastique, Sole Soul, Bobo192, Danwarne, NetBot, Longhair, Feitclub, Func, Rackham, DrYak, Blue Wizard, Elipongo, Cohesion, Angie Y., Nesnad, Kappa, Harcoi, Sasquatch, Kjkolb, Darwinek, Famousdog, Cinnamon~enwiki, Tarun telang~enwiki, Opspin, B0at, Q3aiml, Ardric47, Towel401, Idleguy, Wrs1864, MPerel, Sam Korn, Haham hanuka, Ral315, Hooperbloob, Nsaa, Officiallyover, QuantumEleven, Eje211, Nickfraser, Espoo, Jumbuck, Mithent, Alansohn, JYolkowski, Davidl, ChrisUK, Schnell, Neria, Polarscribe, DenisHowe, Guy Harris, Joolz, Wiki-uk, Lordthees, Atlant, Rd232, Mr Adequate, Andrewpmk, Verdlanco, Ronline, Andrew Gray, Lord Pistachio, Txredcoat, Wikidea, Eagleamn, Yamla, Lectonar, Goldom, Zippanova, Water Bottle, Lightdarkness, Kocio, Mrmiscellanious~enwiki, Monty Dickerson, Chrisjohnson, Mysdaao, Denniss, Malo, Joris Gillis, PeteVerdon, Wtmitchell, Bucephalus, Blancandrin, SidP, Helixblue, Danaman5, ProhibitOnions, Wtshymanski, Jrleighton, Keni~enwiki, Stephan Leeds, Amorymeltzer, Grenavitar, Drat, Red essence, Danthemankhan, LFaraone, BlastOButter42, Versageek, Sleigh, DV8 2XL, SteinbDJ, Alai, Rjhanson54, Algocu, Micdab, HenryLi, Panchurret, Firthy2002, Red dwarf, BerserkerBen, Oleg Alexandrov, Muhgcee, Tom.k, Lafraia~enwiki, Centralman, Dr Gangrene, Novacatz, Flawiki, Thryduulf, Angr, Boothy443, OwenX, Woohookitty, 2004-12-29T22:45Z, LizardWizard, Mindmatrix, Lochaber, Anilocra, Swamp Ig, Pinball22, Localh77, Daniel Case, Uncle G, Polyparadigm, Admrboltz, Pol098, Before My Ken, SP-KP, Byped, JeremyA, Matijap, Tabletop, Privacy, Wikiklrsc, Richardr443, GregorB, MiG, Rchamberlain, Leemeng, Prashanthns, Gimboid13, Liface, Nwahsnwahs018, Static3d, Winged-stone, Palica, NeonGeniuses, ObsidianOrder, Mandarax, Fleetham, MassGalactusUniversum, RichardWeiss, Matilda, Graham87, WBardwin, Deltabeignet, Magister Mathematicae, Chun-hian, SamuraiClinton, Kbdank71, Bunchofgrapes, FreplySpang, Haikupoet, Jasonaj, JIP, RadioActive~enwiki, Miss Pippa, Edison, Rjwilmsi, Sideshowjohn, Phonedude, Quale, Misternuvistor, Commander, Moosh88, Vary, Mjm1964, Hairymon, Tim Eliseo, Quiddity, Tangotango, Sdornan, Bruce1ee, Zpetro, Collard, Mike Peel, Vegaswikian, Kazrak, Ligulem, Sferrier, Boccobrock, Brighterorange, Thekohser, DoubleBlue, MarnetteD, Plasticty, Utuado, Renaissance Man, Yamamoto Ichiro, A Man In Black, SNIyer12, Allen Moore, FlaBot, Ian Pitchford, Sky Harbor, G Clark, Ground Zero, Polaralex, Zeppelin4Life, Jak123, Winhunter, Descent~enwiki, Who, Mathiastck, Nivix, Chanting Fox, Ipats~enwiki, SuperDude115, Abrooks, RexNL, Gurch, RobyWayne, Jrtayloriv, Cbmaster, CoolFox, Snarkibartfast, Lmatt, Webshared, Steve-Baker, Srleffler, OpenToppedBus, Arickp, Dnbosiris, Mrschimpf, Smaley, Party on Aisle 7, Chobot, Kellergraham, Mhking, Stoive, Volatile-Chemical, Cactus.man, Belowzero, Adam J. Sporka, Gwernol, Wjfox2005, Elfguy, The Rambling Man, YurikBot, Wavelength, GMT, Playsta-tionman, Klingoncowboy4, Angus Lepper, Andynormancx, Eraserhead1, Sceptre, Kencaesi, Retodon8, Stormerne, DMahalko, Muchness, Bri-aboru, TheDoober, Splash, Qubeular, DanMS, SpuriousQ, CanadianCaesar, Hydrargyrum, Mythsearcher, Stephenb, Lord Voldemort, Manop, Polluxian, C777, CambridgeBayWeather, Tungsten, Wimt, Pelago, TheMandarin, Rhindle The Red, Terra Green, NawlinWiki, Muntuwandi, Edinborgarstefan, Dat789, Wiki alf, Chick Bowen, Jaxl, Welsh, Joel7687, ZacBowling, Mmccalpin, ONEder Boy, Lone Odessan, Cleared as filed, JDoorjam, Irishguy, Nick, Aaron Brenneman, Dalziel 86, JRG, Anetode, Esther scholle, Brandon, Cholmes75, The Land of Smeg, Matticus78, Night Tracks, PhilipO, Larry laptop, Chal7ds, Rickyboy, Momon526, Obey, William Graham, Hopperlexington, Furball~enwiki, Voidxor, Semperf, Tony1, Alex43223, EEMIV, Brucevdk, M3taphysical, Mysid, Lcmortensen, CDA, DeadEyeArrow, Psy guy, Bota47, Col-inFine, Caspian, Nescio, Bbaumer, Brisvegas, Dv82matt, Tirerim, Navstar, Wknight94, Bob247, Mugunth Kumar, Mütze, BazookaJoe, Saric, Hans Joseph Solbrig, Zero1328, Db9970a, 2over0, Zzuuzz, PTSE, Marketdiamond, Imaninjapirate, Nachoman-au, Teiladnam, Ageekgal, Gim-meahighfive, Closedmouth, VAgentZero, Hrshgn, Fang Aili, Tim Parenti, Pb30, KGasso, David Justin, Tsunaminoai, GraemeL, JoanneB, Alas-dair, Red Jay, Chriswaterguy, Shawnc, Cp33, Anclation~enwiki, Mhenriday, QuillOmega0, Skittle, Garion96, Jonnymoblin, X3210, Bluezy, Mhkay, Arunvijayan, John Broughton, Auroranorth, Roke, DVD R W, Eenu, Ryūkotsusei, Veinor, MacsBug, Vvill, SmackBot, Nick Dillinger, Unschool, Salilm, Hux, Davepape, Iamajpeg, Reedy, Brianyoumans, Tomer yaffe, Prodego, KnowledgeOfSelf, TestPilot, CompuHacker, Marc Lacoste, McGeddon, The Monster, David.Mestel, Unyoyega, Od Mishehu, Thunderboltz, Karmastan, Compay~enwiki, KVDP, Delldot, Chaos-feary, Adammathias, Agentbla, Frymaster, Gjs238, Fnfd, Imzadi1979, BiT, Nil Einne, Edgar181, Mauls, Septegram, SmartGuy Old, Siradia, Aksi great, Richmeister, Unforgettableid, PeterSymonds, Gilliam, Boul22435, Ohnoitsjamie, Choalbaton, OldsVistaCruiser, Andy M. Wang, Sonicandfffan, GeorgeBuchanan, Julie C. Chang, Constan69, Jeffro77, Shrensh, Flurry, Optikos, Kurykh, Spilla, SMP, Cattus, Master of Pup-pets, Thumperward, Raymond arritt, Oli Filth, Ankurjain, Elatanatari, Tree Biting Conspiracy, Jeysaba, Repetition, Joost P. Vermeer, Mdwh, SchfiftyThree, Victorgrigas, Bonaparte, Deli nk, Neo-Jay, Analogue Kid, Jerome Charles Potts, Telecom.portal, Dlohcierekim's sock, ER-obson, Viewfinder, Jfsamper, DHN-bot~enwiki, The Moose, Colonies Chris, Hallenrm, Darth Panda, Firetrap9254, A. B., Andyiou52345, Thief12, Mordantkitten, Brideshead, Sebhaque, Salmar, Zsinj, Decemberster~enwiki, Dethme0w, Can't sleep, clown will eat me, SheeEttin, Vasilken, Volphy, Mitsuhirato, Newcop, DéRahier, Djido, X570, Ioscius, Flibbert, OrphanBot, Onorem, Jennica, Yidisheryid, TheKMan, Rrburke, Homestarmy, Krsont, VMS Mosaic, Addshore, Grover cleveland, Miken2005, Perspective16, Khoikhoi, COMPFUNK2, Jmlk17, Krich, Bindyree, PiMaster3, Solarapex, Jaimie Henry, Cybercobra, Khukri, Nakon, Bsteger, Lubar, Gamgee, LMF5000, Anoopkn, Dread-star, Richard001, Tova86, Rescbr, DylanW, Hgilbert, Weregerbil, SeanAhern, IE, Woodysee, DMacks, Wizardman, Ultraexactzz, Kendrick7, Twir, Salamurai, LeoNomis, Richard0612, Pilotguy, Kukini, Davipo, Bezapt, Keyesc, Ardenn, Ohconfucius, Paul 012, L337p4wn, Lexicon-tra, Tydus Arandor, Snowgrouse, The undertow, SashatoBot, Rory096, Krashlandon, Robomaeyhem, Mksword, AThing, Lester, Harryboyles, Howdoesthiswo, BrownHairedGirl, Stewie814, Zero10one, Kuru, Celluser, JackLumber, Jidanni, Vitall, Roarke, Scientizzle, Shaliron, Soptep, Mwboyer, Freewol, J 1982, Heimstern, Nobodyinpart, Disavian, Calum MacÙisdean, CPMcE, Rohan Lean, Tennis Dynamite, Wibbble, JohnI, Soumyasch, Sir Nicholas de Mimsy-Porpington, Shlomke, Tony Corsini, Slinga, GCW50, MidnightSwinga, Accurizer, Goodnightmush, ManiF, AMac2002, Gregorydavid, Jaywubba1887, Aleenf1, Rawmustard, Anand Karia, 041744, Ckatz, Cortezz~enwiki, CyrilB, 16@r, Ex nihil, Sinaerem, JHunterJ, MarkSutton, Stupid Corn, Andypandy.UK, Slakr, Karn-b, Drumersrule, Beetstra, LuYiSi, Milesdowsett, Mr Stephen, Dick-lyon, Mariersteve, Larrymcp, Waggers, Dammit, Felixluo, TastyPoutine, Synergism, Tuspm, Goldline~enwiki, Gizmoleeds, Mintchocolatebear, MTSbot~enwiki, H, PSUMark2006, Haveronjones, Agent 86, Dl2000, Andreworkney, GorillazFanAdam, Hu12, Stephen B Streater, Hetar, BranStark, OnBeyondZebrax, Chuck 266, Fan-1967, Iridescent, Ft93110, Stangoldsmith, Joseph Solis in Australia, Kernow, Newone, JHP, Casull, Cradle, The nibbmeister, Igoldste, Cls14, QcRef87, RekishiEJ, Beno1000, Esurnir, Estrategy, EurowikiJ, Svego~enwiki, Mr Chuckles, FIshstick, Linkspamremover, Illyria05, Rayoflight278, Eluchil404, Tawkerbot2, Hogibear, I3lizzard, Simplyagro, Chris55, Flubeca, Absolut-Dan, Lahiru k, JohnTechnologist, Rpb161, SkyWalker, J Milburn, JForget, RSido, Vega84, CmdrObot, Ale jrb, Cxw, Top Cat, Dread Specter, Wafulz, Zarex, Ric36, Jorcoga, Comrade42, KyraVixen, Cellphones and Pagers, The White Cat, Dgw, Splenius, Kai360uk, Priceytom, Small-pond, GargoyleMT, Imaginationac, MarsRover, WeggeBot, Kalemika, Moreschi, Smoove Z, Jerressy44, Ken Gallager, Joostjodel, Trex005, Omnicog, Vanished user fj0390923roktg4tlkm2pkd, Nilfanion, Votingontheinternet, AndrewHowse, AMFilmsInc, Pit-yacker, Grenno, Cyde-bot, Abcde123456, Dynamic1, Mashby, MVimislik, Neurocistance, Shritwod, Stephanbim~enwiki, Garyruskin, Gogo Dodo, Xxhopingtearsxx, Corpx, Swat828, Mglickman, Coolguy22468, Myscrnnm, Pascal.Tesson, Scott14, Simmysimsim, Dancter, LaserBeams, Dr.enh, Evelynchai, B, Tawkerbot4, Codetiger, Energetic is francine@yahoo.com, DumbBOT, Chrislk02, Starionwolf, Shoobe01, Pokeman, In Defense of the Artist, Bpadinha, Legis, Kozuch, Ward3001, Abtract, Arcayne, Omicronpersei8, Vanished User jdksfajlasd, Zalgo, Michaelorgan, PamD, Kuang Eleven, Xwas, .:.Critical.:., EvocativeIntrigue, Thijs!bot, Epbr123, Andy Eng, Dr.Bhatta, Wikid77, Ashlie123, Lov ya, Qwyrxian, Lfrench, Daniel, HappyInGeneral, Josephbrophy, CompC, Keraunos, Nachdenklich, Mojo Hand, Marek69, Electron9, Peace01234, AlexanderM, Over-ridex, Doyley, Second Quantization, Z10x, Tellyaddict, Cool Blue, Sinn, LG4761, Hcobb, AgentPeppermint, InfernalPanda, Zachary, DaveJ7, Blathnaid, Srose, Mobilekick, DeusMP, Lajsikonik, SusanLesch, Dawnseeker2000, Eiffelle, AlefZet, Escarbot, Pie Man 360, Eleuther, Lach-lanA, Hmrox, KrakatoaKatie, AntiVandalBot, Rahularora1985, Majorly, Joejoew97, Gioto, Luna Santin, Caledones, Sirishar, Stile4aly, Ope-lio, Alois.Daniel, CZmarlin, Fru1tbat, Apiya27, 29productions, TheBlueFox, Prolog, Benclinch, Autocracy, Khin007, Almondwine, Tmop-kisn, Tjmayerinsf, Postlewaight, Arghlookamonkey, Caffeinatedblog, Kristoferb, Povins, Sehsuan, Ankushj, Ran4, Alphachimpbot, Kzaral, Leevclarke, Lfstevens, A*A*, Canadian-Bacon, Kniwor, Golgofrinchian, Res2216firestar, Kariteh, HanzoHattori, JAnDbot, Key-global, Har-ryzilber, Bobvila2, BenjaminGittins, MER-C, Kprateek88, Wesborland, Sonicsuns, Ericoides, Arch dude, Blood Red Sandman, Cleanupman, Dhp~enwiki, Grant Gussie, Sosh, Andonic, Roleplayer, Hut 8.5, Dave101, Greensburger, Dricherby, PhilKnight, AOL account, Rothorpe, Joneyf1, GoodDamon, Madhive, Y2kcrazyjoker4, LittleOldMe, SiobhanHansa, Acroterion, Yosh3000, MaxPont, Pedro, Whatever626, Bong-warrior, VoABot II, Peas and corn, Dannyc77, AuburnPilot, Jpmaurya, HorseloverFat, JamesBWatson, Adrian122, Think outside the box, BobTheMad, Cadsuane Melaidhrin, Pugetbill, Cumiskey192, Trottsky, Mphung, Euhedral, Nyttend, Paul Haymon, Dendrolimus, Froid, Nick Cooper, MadeinIndia, Bubba hotep, Catgut, MobileMistress, Have hat, will travel, Indon, Times10, Dhurowitz, Acornwithwings, Frijole, Arch-Stanton69, Mittosi, Thedreamdied, Robotman1974, Stoneice02, 28421u2232nfenfcenc, Badie, TerryChen, Allstarecho, Kingutd, Tswsl1989, Netboyak, Gomm, Marinebro0306, V 1993, Glen, Chris G, DerHexer, GermanX, Esanchez7587, Streetsk8ta4life, Cool.x, Patstuart, Flami72, Tidus187, Stewartpalmer, Khr0n0s, Gjd001, Stephenchou0722, Onlynokia, S3000, Alx 91, MartinBot, Dennisthe2, Bentleywannabe, PaulLev, Andycole, STBot, Gasheadsteve, Musicmelodygirl, Arjun01, Tvoz, Solcuerda, Umeshunni, Jim.henderson, Brogman, Rettetast, Bissinger, Andsam, Jonathan Hall, Tskam1, Burnedthru, Kostisl, Mycroft7, CommonsDelinker, Duncanelliott, Nono64, AgarwalSumeet, PrestonH, Siliconov, Antb, Jeroldc, Slugger, HarZim, J.delanoy, Captain panda, Martinf hk, Pharaoh of the Wizards, Mange01, MoiraMoira, Trusilver, Darin-0, Rgoodermote, Chinese goods, Neutron Jack, Quiksilver11709, Ali, Bogey97, Redtestarossa, All Is One, SimpsonDG, Jesant13, Ginsengbomb, Boris Allen, Extransit, TrueCRaysball, Geoweb54, WarthogDemon, SedireX, Thaurisil, NerdyNSK, Ian.thomson, Scottrb, The-gr8, Davidprior, Acalamari, Ivanam, Aestiva, Cazanoma, DanielEng, Kingj123, WikiBone, It Is Me Here, Elkost, Monty54, Bot-Schafter, Katalaveno, Markseth, Atomichippo, Hometownmallonline, DarkFalls, McSly, Ignatzmice, Apurv1980, Tetonca, Mikael Häggström, Skier Dude, Ahodacsek, Beeva24, AntiSpamBot, (jarbarf), TomasBat, Vaughanemery, Michaellubelle, Btd, SJP, Carewser, Cobi, Mobile Cell Phone Forensics, Touch Of Light, ThinkBlue, Tanaats, Joeykry, Toby lodge, Potatoswatter, Grmarkam, Schwal, Deshawnjon, Cmichael, Andy Farrell, Juliancolton, Cometstyles, RB972, Yosarian~enwiki, Greatestrowerever, Jamesontai, Noctuidae, Crummie, Shawt-NurRhd, Danield101, Gtg204y, TWCarlson, Culdesacjungle, SquishyTheQueeniac, Useight, StoptheDatabaseState, TheNewPhobia, CardinalDan, CrZTgR, Idioma-bot, Funandtrvl, DecadeMan, Jc7919, Echosmoke, Black Kite, Lights, Buttin, X!, Flikes123, Dragonvulture, LeeColleton, Hellno2, Shiggity, The stripy penguine, Hammersoft, VolkovBot, Traf22, CWii, Science4sail, SimplyCellPhone, Leebo, Jeff G., Indubitably, Dmaonk, Mathteacher1729, Bovineboy2008, Soliloquial, Helplessstar88207~enwiki, Aesopos, Gracemusante, Cacolantern, Philip Trueman, WalrusMan118, RPlunk2853, FluffyWyld2, Trenwith, Martinevans123, Isaac Sanolnacov, Drunkenmonkey, Blendus, RiSeLing, TXiKiBoT, Prieplauka, Newtown11, Eve Hall, SethFreeman, Thepulse2007, Mettler, Aghibner, SK hockey fan, Graavy, Anonymous Dissident, Funky Monkey 2000, Kishonti~enwiki, Tangerineduel, WikipedianYknOK, Qxz, Piperh, RMWLLC, Warrush, Anna Lincoln, Aeharding, Lradrama, Melsaran, Kovianyo, CanOfWorms, Bagle, Abdullais4u, Canaima, LeaveSleaves, Edgar187, Raryel, Darkkaangel, Tim9798, PDFbot, Cremepuff222, Maxim, ARUNKUMAR P.R, C-M, Bkbroiler, Tri400, Shifter95, Sses401, Vladsinger, Jensgram, Aron.Foster, Graymornings, Poopyunderwear1, Gracenotes' left sock, Synthebot, Altermike, Saramarie8789, Falcon8765, Hg5131, Richtom80, RMW42, Vchimpanzee, MyronAub, Sesshomaru, Unused0030, Runarreistrup, Amalia07, Rboxer, Agape25, Superdude444, Doc James, AlleborgoBot, Nagy, Symane, PGWG, Pc9889, Cheesecake13, PRChinaMan, Wwwwhite, Fearthisname, Jackky, Red, D. Recorder, Hmwith, Redoggy101, Marioblooper, Kbrose, DenverF666, Jimmayoy007, The Random Editor, Biscuittin, 3e solutions, Darkieboy236, Markomiha, SieBot, Froztbyte, Whiskey in the Jar, Smlowe5, Celeste67, Thennessey8910, Anhimgr8, Augustus Rookwood, YonaBot, Scarian, Jevo90, Luboogers25, Malcolmxl5, Ellbeecee, Loveshoes, Ashayh, Ypps~enwiki, Hertz1888, Jack Merridew, Iveschavira, Luisa Yeom, Da Joe, Caltas, Karlson753, Twinkler4, Shochin, DIAMOND LOVE2007, RJaguar3, Lcfuentes, CurranH, Feoray, Gravitan, LeadSongDog, Cbird09, Infodriveway, Bhagwatkumar, Nummer29, Wheelspins13, Dllew1, Blake3522, Archers princess, Jrfireboy2, Cocomaxwater, Chromaticity, Tiptoety, Radon210, Smsmasters, Larek, TheThingy, Glenibaby06, Stumbler7, Nopetro, Eamclellan, Exrev, Jnottingham13, JetLover, Undead Herle King, Aureliusweb, Jimthing, Jimblobodob, Guildwarsplayar, Istaro, Oxymoron83, Kkgorykid, Sammygooders, Faradayplank, AngelOfSadness, Physik, Editor91, Peter k john, Lightmouse, Tomi T Ahonen, Kd24, Macy, Sacon, Wrmscomet, Ooza, Spitfire19, Schmidty247, Correogsk, LonelyMarble, StaticGull, Martycell, Cyfal, Kdevin, Mygerardromance, Realm of Shadows, Thedarkcookie, Immy92, Vig vimarsh, Dust Filter, Altzinn, MichaelIvan, Petete333, Klaus100, Kelan19, Dabomb87, Superbeecat, Florentino floro, Cutiepie 10129, Princess habibu, Daveruzius, Adnanp, C0nanPayne, Tomahiv, Vonones, Perspective Vortex, Explicit, Mcarrieri, Falcon-eagle2007, Anthony060708, Cactuscake, Faithlessthewonderboy, Loren.wilton, Sfan00 IMG, Tanvir Ahmmed, Elassint, ClueBot, Admiral Norton, Schaea, Pressforaction, Cpljwlusmc, Stinkehund, PipepBot, Snigbrook, Badger Drink, Foxj, LondonBVE2, The Thing That Should Not Be, Cellfire1, Gpermant, TherionPhusikos, Rjd0060, Tsomas214, Cambrasa, EoGuy, Royalgam, Figarema, P0mbal, Frmorrison, SecretDisc, Phone2phonee, Klabber, CounterVandalismBot, Leodmacleod, Shaliya waya, Niceguyedc, Blanchardb, Harland1, Argonistic professor, Klabber1, Xjohnjohnx, Bob bobato, Neverquick, Volleyballspider, Danaeb, Michaelmoran, Namazu-tron, Glaurio2000, Amirul2008, Mmark089, MrTobe, Silverbackwireless, Mr. Laser Beam, Thewhoissuperb24, AcademicallyIntelligent, D10025g, 123456789abcdefghi, DragonBot, SteveRamone, SapientiaSativa, Pc on wheels, Gimli007~enwiki, Excirial, KihakuNoSenshi, Iloveulikeafatkidlovescake, Congolese fufu, Fireflyfan1, Ashirv, Dan373, Rodgersmas214, Resoru, Smartfunda, Mas214Kapinga, Daymas214, John Nevard, ZHUMAS214, Jeffwashere, Smallmas214, Estirabot, Lartoven, TUMAS214, Lunchscale, Jotterbot, Chirag Patil, Okiefromokla, Promethean, Pokespot, GufasBorgz7, JamieS93, Gbvjcczxhb, RayquazaDialgaWeird2210, Mgw89, Dekisugi, Dnaphd, Gundersen53, Queenmegmeg, Nukeless, Careida, Nedim.sh, Sallicio, Ottawa4ever, Polly, ChrisHodgesUK, Swingout, Orionrulesallpeeps, Sudhir.rao, Thingg, Error −128, Quinn111, Aitias, Volodimir~enwiki, Versus22, Teleomatic, Dana boomer, Mshadowsbabywoo, Xman4sho, Djk3, Katanada, David Chiam, Chuuplo, Johnuniq, SoxBot III, Apparition11, SF007, ClanCC, Ginbot86, DumZiBoT, InternetMeme, Loo4, Dairyqueen8, XLinkBot, Mattthistle, Emmette Hernandez Coleman, Gwandoya, Rror, Duncan, Mitch Ames, SilvonenBot, NellieBly, Ericloewe, Sawickin, Sweetpoet, Noctibus, Pcpersons, NHJG, Zodon, Cooljc~enwiki, Catgirl, Sk8erking85, Sassygirl101leahu, Supercool277, Websi7, Thatguyflint, Kbdankbot, Addbot, Speer320, Schesnais, Vandalism bot1, Jcfan710, Witysmartone, Bluejm2, Willking1979, AVand, Some jerk on the Internet, DOI bot, Alex1city, Ente75, Kiy765630, Fgnievinski, Pjjafc, Orangesbob, Nibbles249, Pieman196, Monkeysocks2, Ja2ck2ie2, Mohamedhp, Battleboom13, Naughtyshakil, CanadianLinuxUser, Rj1200, NjardarBot, Ka Faraq Gatri, Cst17, MrOllie, Austinnielson, Belmond, CarsracBot, Thom443, Ccacsmss, Lalalosurr, Glane23, DoraXplorer, Mathu321, Bloopasareverykool, Buddha24, Favonian, Jasper Deng, West.andrew.g, Tyw7, Livni, Ssseeeaaan, Deathspawn the suicidal, Numbo3-bot, Wikinoob123, Theking17825, Evildeathmath, Cylover, Sinadoom, Tide rolls, Ghustaff0809010, Verbal, Krano, Nuberger13, Aadieu, QuadrivialMind, Gail, Wireless friend, Abracadabration, Narayan, Jackelfive, Chaldor, Luckas-bot, Hikaru-Tora, Yobot, Fonboozles, Walter.Geo, Geroldorules69ers, Tohd8BohaithuGh1, JSimmonz, Cflm001, Specious, Donfbreed, II MusLiM HyBRiD II, GOR42, KamikazeBot, IW.HG, South Bay, Tempodivalse, OregonD00d, Retro00064, Backslash Forwardslash, AnomieBOT, Bctwriter, Lilduff2008, 1exec1, IRP, Galoubet, Pyrrhus16, ISquishy, JackieBot, Majikboom, Solidsandie, Flewis, Bluerasberry, Limideen, B2031919, Rtyq2, Crockie422, Citation bot, IRKAIN2, Lisamcghee87, GACHealth, Syberiyxx, Patentideas, Glevinso, Frankenpuppy, Neurolysis, ArthurBot, Quebec99, Bobbyjoe101, Speedstick76, Rosenblattl, Xqbot, ManningBartlett, Bjorn Elenfors, Transity, Dockfish, Addihockey10, Capricorn42, Termininja, Robot85, Khajidha, GenQuest, TheWeakWilled, Repsrule, GrandKokla, JascalX, □□□, Tomwsulcer, Connshearer, Handshaines, Karthik6129, Wiki2contrib, Shahzad11, GrouchoBot, Burstorange, Waffleboy666, 0wnag34life, Ute in DC, Wizardist, Kopieto, Mark Schierbecker, SassoBot, Karghazini, Southberry, Kcdtsg, KMoore175, Catstuffer1, Kennny27, Yoganate79, Eckerslike, Connerc 11, Lyle-Howard, Shadowjams, Kyle Hardgrave, Bluesoju, Taka76, Klapouchy, FreeKnowledgeCreator, GliderMaven, FrescoBot, Akuvar, Kat081685, Chrsschm, Tobby72, GEBStgo, Tiramisoo, JuniperisCommunis, Varundbest10, Troglo, Austria156, DivineAlpha, Wireless Keyboard, HamburgerRadio, Avani089, Elvisforpresident, Citation bot 1, Zhwn24, Tonyupward, Ryannie1991, Rsolero, Nokia ST, Cpflieger, I dream of horses, Seals9889, Robert A. Maxwell, Chatfecter, The Arbiter, Toomuchcash, Angstorm, Dneubert, Tinton5, Cooltwig, PSUdesigner, Vincenzo 90, Jamesinderbyshire, Phoenix7777, Sajalkdas, Jirka.h23, FoxBot, Mjs1991, VEO15, TobeBot, Yunshui, DriveMySol, Lionslayer, Lotje, Ladiesman2215, Extra999, Olliecracknell, Aoidh, David Hedlund, Crysb, Lambanog, Trinary M01, Jamiespinks, Iloveyourfaceman, MegaSloth, FelixtheMagnificent, H.ehsaan, DARTH SIDIOUS 2, Northwestern guy, Renault555, Horshamknowsbest, Maryjaneadams, Lacastrian, Meeghaman, RjwilmsiBot, Anotherskinnykid, Scoreymjp, Vtstarin, Bokorember, NameIsRon, Ripchip Bot, Carnell robison, Pioneer valley, Realglobalist, Carolynlowe, Caster33, Lopifalko, Rubinl88, Esibun, Powerkeys, Chapy26, BobbyChristmas, Zaeriuraschi, Glamour girl1994,

Bo789789, Becritical, Spidey9995, Stinkyindian, Northwesternep15, EmausBot, John of Reading, Repiceman89, WikitanvirBot, Ghostofnemo, Ever388, Emmalewis1, Karagirl25, Rusty Sheklelfurd, Dewritech, Qjawls2030, Modou 1990, Yt95, GoingBatty, Cicely5, G&CP, Zachanders, Wikipelli, Americanhaute, Evanh2008, Kkm010, Listmeister, Udvarias, Acategory, Jack Sebastian, Arunj 001, Idh0854, Samgj127, Pyro721, Unreal7, Alex Neman, Hamiltha, Palosirkka, Gsarwa, Rangoon11, ChuispastonBot, Wakebrdkid, Targaryen, DJDunsie, Ksmit139, Georgy90, Woolfy123, Diamondland, ResearchRave, Mikhail Ryazanov, Emma23 K, Jwhimmelspach, ClueBot NG, Jnorton7558, W.Kaleem, Matthiaspaul, Gilderien, Daveduv, David O. Johnson, Singhmahendra20, Theimmaculatechemist, DieSwartzPunkt, Mesoderm, ScottSteiner, Ivan345garcia, Yev121, Abcd888, Nigs12345, Zackaback, Chacha15, Funkymonkey1997, Aurelie Branchereau-Giles, Joshb094, Joshb09444, Theopolisme, Mobileready, Utgard~enwiki, Helpful Pixie Bot, GGink, FtDesoto, Julietlewis, Roshanamila, Ceekee, පසිඳු කාවින්ද, Webzoneme123, Phonenotebook, Nomam123123123, Skaldragon, Gob Lofa, Tyagirl100, Pine, Smart1954, Kndimov, Northamerica1000, RobLandau, AvocatoBot, Davidiad, Bobultranerd, Architect101, NNU-01-05100137, Rsamahamed, Ddavid2005, Shirudo, Russianamerican1, Cynival, BattyBot, Bagoto, RichardMills65, SchreibStang, Brimur05, Khazar2, Froekenjul, Harpsichord246, MilanAD, HuntersMoon22, BrightStarSky, Katiewiltshire, Dexbot, Aditya Mahar, Blobbie244, CuriousMind01, KarinaJ son, Lugia2453, Kskhh, Kevin12xd, Reatlas, Gabby Merger, Karthikeyankc, Polytope4d, Seqqis, Zalunardo8, Ryenocerous, Sahildhiman27091999, Andrewlyly, Thevideodrome, Spyglasses, Melody Lavender, Gokul.gk7, Saiful7, MameTozhio, Publiususa, Meteor sandwich yum, Lagoset, CyHack, Monkbot, Soloism, Dannywong1190, Filedelinkerbot, Qwertyxp2000, Amarkowitz1, Bpmeller, Wiki.wonder.56, GeorginaMat, Maodit, Kaufmanitay, Morlvi, Junaid sipra, KasparBot and Anonymous: 2609

- **Mobile phone radiation and health** *Source:* https://en.wikipedia.org/wiki/Mobile_phone_radiation_and_health?oldid=686977821 *Contributors:* Ghakko, Rsabbatini, Edward, Modster, DopefishJustin, Tompagenet, CesarB, Egil, Timwi, Dragons flight, Furrykef, Saltine, Floydian, Dittaeva, Pingveno, Meelar, Auric, Sunray, Aetheling, Giftlite, Bovlb, Solipsist, Albany45, Carlos-alberto-teixeira, Erich gasboy, Andycjp, Beland, CaribDigita, FrozenUmbrella, Esperant, Qui1che, CALR, Arcataroger, Rich Farmbrough, Vsmith, LeeHunter, ESkog, Kaisershatner, J-Star, RoyBoy, Lyght, Lorem Ipsum~enwiki, Bobo192, Enric Naval, Shenme, Cmdrjameson, Brim, JeR, Slicky, Towel401, John Fader, Alansohn, ChrisUK, Andrewpmk, Wouterstomp, Avenue, Stillnotelf, Velella, Omniscientist, Tony Sidaway, Seec77, Anonymous3190, Rgrig, BerserkerBen, Woohookitty, Mindmatrix, Awostrack, Barrylb, Benbest, Bonus Onus, JeremyA, GregorB, ObsidianOrder, Mandarax, BD2412, Elvey, David Levy, Rjwilmsi, Nightscream, DeadlyAssassin, Tangotango, Vegaswikian, Keimzelle, Tedd, JohnGH, Ground Zero, Nihiltres, AlastairR, Jrtayloriv, Ahunt, DVdm, Bgwhite, Wavelength, Hairy Dude, Bhny, Pigman, Chris Capoccia, Hydrargyrum, Dotancohen, Bachrach44, Badagnani, Dureo, Bucketsofg, Jhirvi, M3taphysical, Mysid, DeadEyeArrow, Kkmurray, User27091, Angusj, Hans Joseph Solbrig, American2, 2over0, Chase me ladies, I'm the Cavalry, Closedmouth, Trendall, Petri Krohn, Mossig, NeilN, Zvika, Tom Morris, SmackBot, Roger Davies, Nmshyam, Tomer yaffe, KnowledgeOfSelf, TestPilot, CyclePat, C.Fred, DWaterson, Ifnord, Delldot, Ahmadr, Lainagier, Quinkysan, Commander Keane bot, Unforgettableid, Gilliam, Ohnoitsjamie, Kmarinas86, Chris the speller, Bluebot, Keegan, Magical shooter, Thumperward, Joel.Gilmore, ERobson, Colonies Chris, A. B., Dethme0w, Can't sleep, clown will eat me, Krsont, Kyle sb, MitchellShnier, Cybercobra, Nakon, DoubleAW, Brainyiscool, Lcarscad, Kendrick7, Parrot of Doom, Ligulembot, Mchavez, ArglebargleIV, AThing, KLLvr283, Jaganath, Ksn, Tktktk, Mgiganteus1, Grumpyyoungman01, Topazg, Meco, Waggers, Arathalion, Daviddaniel37, Jcbutler, DabMachine, Olivierd, Iridescent, Alessandro57, RekishiEJ, Noodlez84, Hyperman 42, Tawkerbot2, Gveret Tered, CmdrObot, Green caterpillar, FlyingToaster, Tex, BenGriffiths, Cydebot, Gogo Dodo, JFreeman, Christian75, Mtpaley, Omicronpersei8, Pipatron, Talgalili, Epbr123, Lfrench, Grand Dizzy, Hazmat2, N5iln, Mojo Hand, Headbomb, Bobblehead, Electron9, PaperTruths, A2X14, Big Bird, Viralmemesis, SvenAERTS, Greenpand, Obiwankenobi, Tjmayerinsf, Darklilac, DarthShrine, Steelpillow, Mikenorton, MER-C, Ph.eyes, Hello32020, Ipso-De-Facto, Charles01, KyleAndMelissa22, VoABot II, JamesBWatson, Pixel ;-), Twsx, Cgingold, A3nm, DerHexer, Garik 11, Hbent, DGG, Yobol, MartinBot, Bewareircd, Axlq, Sm8900, R'n'B, Nono64, Ash, Ewan dunbar, J.delanoy, The dark lord trombonator, Mojodaddy, Trusilver, AstroHurricane001, Katalaveno, Mahewa, Silas S. Brown, Mikael Häggström, DadaNeem, Gtg204y, Inter16, Useight, Levydav, Taprootdancer, Shuriken00, Caspian blue, VolkovBot, ABF, Ericdn, Mrh30, Jasonsaurusrex, Philip Trueman, Martinevans123, Prd34, Msdaif, Joe2832, Anonymous Dissident, Darkanius, Davemc50, Qxz, Anna Lincoln, Anathea, Sciencewatcher, Onore Baka Sama, Madhero88, Stoilis, @pple, Femputer, GlassFET, Harrolegg, Ronmore, Emma li mk, Kingbusch, The Realms of Gold, Hmwith, AS, Pallab1234, Hertz1888, Yintan, Srushe, Grundle2600, Happysailor, Trustotherplluijj, Allmightyduck, Oxymoron83, Faradayplank, Harry~enwiki, Vegasciencetrust, Svick, MadmanBot, Garrettw87, Superbeecat, Pinkadelica, Wdwd, ClueBot, Artichoker, The Thing That Should Not Be, ImperfectlyInformed, Tabby2K7, Jumacdon, Uncle Milty, Niceguyedc, LeoFrank, Excirial, Crywalt, Leonard^Bloom, Ph33NIkZ, NuclearWarfare, Thingg, Jonverve, SDY, RetiredUser58, Kozzz, DumZiBoT, InternetMeme, Make poverty matty, RDOlivaw, XLinkBot, DrEightyEight, Avoided, Alexius08, SchwarzeMelancholie, Addbot, Grayfell, NotThatJamesBrown, DOI bot, Jojhutton, PleaseInsertGirder, Otisjimmy1, GeoffreyBanks, Franmars, Snemana, Fluffernutter, Angelia2041, LaaknorBot, McSaucePaste, BExelby, 22yearswothanks, Farflungwaffle, Oldmountains, Tide rolls, Verbal, Lightbot, Pensees, Guyonthesubway, Gail, LuK3, Telial, Ben Ben, Luckas-bot, Yobot, Fraggle81, Kilom691, Billywalker, KamikazeBot, SwisterTwister, Caleb Rentpayer, Wikid777, Retro00064, Raimundo Pastor, AnomieBOT, Ormers, Ubergeekguy, Flewis, Materialscientist, Citation bot, Roux-HG, Williamsburgland, LovesMacs, Quebec99, Sionus, Kyeller, Nasnema, Articjuice, Skarl the Drummer, J04n, Kylelovesyou, Ajax151, Shadowjams, StoneProphet, Robecko, Jc3s5h, Mykolanovik, Seleonov, יובל מדר, Citation bot 1, Vicenarian, Notedgrant, 05wrightc, Jonesey95, M for Molecule, Mrblablabla, Île flottante, Meaghan, Full-date unlinking bot, Trappist the monk, Yunshui, Duncwilson, ItsZippy, Kmw2700, Viktor Laszlo, David Hedlund, Fiberglass Monkey, Diannaa, Ivanvector, User8253, DaveACIM, RjwilmsiBot, Jojo129~enwiki, Torontokid2006, Regancy42, Hadolaven, Becritical, Timtempleton, Tallungs, Wadey1997, ScottyBerg, Dewritech, Stinky cheese footface, Weirdal176, Waveban, Daniel G US, GoingBatty, RenamedUser01302013, Artiomjar, Slightsmile, Scgtrp, ZéroBot, Johnson aj, DanielJohn656, Arunj 001, Hazard-SJ, Can You Prove That You're Human, H3llBot, SporkBot, FinalRapture, Thine Antique Pen, IGeMiNix, L Kensington, Danmuz, Mattikins, Happiness366, GrayFullbuster, Angelnow3, DASHBotAV, Radiohead1989, 28bot, Rocketrod1960, Kinkreet, Petrb, ClueBot NG, RaptorHunter, Widr, Spfoster, Jrassmussen2010, Isa58, Bouji, Helpful Pixie Bot, Anentiresleeve, Krkwrgr, Calabe1992, BG19bot, Pine, Slinc123, Chgsonx, HENDOAGAIN, Vicky fobel, Neøn, Howbig~enwiki, Snow Blizzard, Asto2121, Goush.ahmad, Hamish59, 6Dayz2s33, Riley Huntley, Pratyya Ghosh, Mrt3366, JMercer2012, Catie xx, Drfrankv, Gus10011, Devilzay, Prkaru0007, BigJolly9, Mogism, Cerabot~enwiki, UcheD43, Frosty, Graphium, Little green rosetta, Jamesmartin01, Perrinald, BruceLevesque, RetiredHelp, Jodosma, Franoprl, Everymorning, Karamura123, Juphill, DavidLeighEllis, Comp.arch, Shadehz, Trtehp35, Jianhui67, Hindydrillick, Ikabobjoe28, Marcusknight23, Jameskirk, JaconaFrere, Monkbot, JohnWhoMeansWell, VG31-irl, RobertBDurham, AKB211, Spumuq, Bklunde, Pranavpina, Pommapple, Raeggggaeshark, Braedster10, RA5957, KasparBot, Mandeepsinghdua, AmCharlie, Haleyroughton, Jaramioa, Glory of Space, Perrymasonfan and Anonymous: 711

- **Bioinitiative Report** *Source:* https://en.wikipedia.org/wiki/Bioinitiative_Report?oldid=656943419 *Contributors:* William M. Connolley, Qui1che, Giraffedata, Ground Zero, Srleffler, Tone, Wavelength, 2over0, FyzixFighter, SmackBot, DMacks, Ohconfucius, Topazg, AubreyEllen-

Shomo, Rettetast, Keith D, Funandtrvl, ClueBot, Kozzz, Addbot, CaneryMBurns, Verbal, Pensees, Yobot, AnomieBOT, Citation bot, Stanislao Avogadro, Full-date unlinking bot, NottsStudent09, Augratin, Alasdairmp, Hadolaven, EmausBot, Bio201001, H3llBot, ChuispastonBot, Clue-Bot NG, Blessingsncheers, Liam987, TopGarbageCollector, Megabytehertz, Monkbot and Anonymous: 11

- **Dielectric heating** *Source:* https://en.wikipedia.org/wiki/Dielectric_heating?oldid=684393602 *Contributors:* Itai, Mcapdevila, Rchandra, Beland, Sparkgap, Kjkolb, Ashley Pomeroy, Wtshymanski, Nadovich, Rjwilmsi, Wragge, FlaBot, Gurch, Cryptic, Salsb, Mossig, SmackBot, Chris the speller, Thumperward, Sbharris, Squibman, Pulu, BullRangifer, Evlekis, Aeluwas, Bendzh, Kvng, Andreas Rejbrand, Robinavery, Gatortpk, MaxEnt, Cydebot, Christian75, Surly Dwarf, Thijs!bot, Barticus88, AntiVandalBot, Magioladitis, Faizhaider, JaGa, Mikael Häggström, Bonadea, Almazi, MaltaGC, Oxymoron83, Explicit, De728631, ClueBot, PixelBot, Rhododendrites, Dthomsen8, Mitch Ames, Subversive.sound, Addbot, Bwrs, Tide rolls, Verbal, Yobot, Carlingkirk, Vini 17bot5, AnomieBOT, Götz, Materialscientist, Citation bot, Bhushaiah, Remotelysensed, A8UDI, YourMotherLikesSnape, Orenburg1, J. in Jerusalem, Hhhippo, H3llBot, Staszek Lem, Thouny, Donner60, Rocketrod1960, Helpful Pixie Bot, BG19bot, MrBill3, Iridule, Sojovictor, Cephas Borg, Adirlanz, Djtfoy, Blart versenwald and Anonymous: 61

- **Electromagnetic hypersensitivity** *Source:* https://en.wikipedia.org/wiki/Electromagnetic_hypersensitivity?oldid=684700901 *Contributors:* The Anome, Gabbe, Karada, Skysmith, Charles Matthews, Timwi, IceKarma, Silvonen, E23~enwiki, Floydian, Xanzzibar, Acampbell70, Solipsist, DragonflySixtyseven, Qui1che, Rich Farmbrough, D-Notice, Dbachmann, Zeality, Espoo, Anthony Appleyard, Titanium Dragon, Bobrayner, Tuur, Rjwilmsi, Ligulem, Kolbasz, Kayman1uk, Bhny, Motmot, Robertvan1, Lipothymia, Jhirvi, Mysid, Kkmurray, 2over0, Abune, JQF, Mossig, SmackBot, Thorseth, DWaterson, Diegotorquemada, Peter Isotalo, Chris the speller, Pieter Kuiper, Just plain Bill, Acdx, JzG, JoshuaZ, Topazg, Trey56, Meco, AdultSwim, Arathalion, Alessandro57, Hyperman 42, Thricecube, Tanthalas39, Alan1507, Steelpillow, Hut 8.5, Mim007, Trekie9001, JamesBWatson, Ubershmekel, WhatamIdoing, Jimjamjak, Catpig, Wshallwshall, Sm8900, Tskam1, Mange01, Tiyoringo, Kyle the bot, Sciencewatcher, Noformation, GlassFET, Locke9k, Doc James, Jehorn, MaryLed, Boris999~enwiki, SieBot, Rod Read, Caulde, Phe-bot, Toddst1, Alexbrn, Jagra, Lightmouse, Kd3qc, Twinsday, SlackerMom, ClueBot, CiudadanoGlobal, Catpigg, ImperfectlyInformed, Unprovoked, Yoman82, RetiredUser58, Perchy22, Robodoggy, Anubad95, RDOlivaw, Autumn snake, Benboy00, Partyoffive, Addbot, Proofreader77, NotThatJamesBrown, DOI bot, Simonm223, GeoffreyBanks, SpellingBot, Fluffernutter, Looie496, CaneryMBurns, Verbal, Pensees, Bultro, Legobot, Yobot, DavidWestlake, ScienceMind, AnomieBOT, Citation bot, Mbruck, Xqbot, OlmoGentile, Lennywilson, Igor Mortis~enwiki, Jujuvee, Smallman12q, NakedTruthReport, SolarSauna, Denver26~enwiki, Citation bot 1, Pinethicket, Dront, Trappist the monk, Kmw2700, Warrah, Chronulator, Alasdairmp, EmausBot, Cricobr, Timtempleton, Clark42, DavidLitle, ZéroBot, H3llBot, Donner60, Bluplanet, Teapeat, ClueBot NG, Robertwilliams232323, Jbrandon2012, Widr, Isa58, Helpful Pixie Bot, Itzuvit, BG19bot, Filipegarcia, Badon, Saere, Pm57a, GreggoTw, MrBill3, Bobheide, APerson, Melissarama, Agn106, Daniel Helman, Everymorning, Luxure, Marcusknight23, Stamptrader, Rosesollere, Ken Kozminski, Jasonlee8985, Monkbot, Ehskid, Kylesmith55, Lucianotorredo, Jerodlycett, KasparBot, Panewithholder and Anonymous: 138

- **Electromagnetic radiation** *Source:* https://en.wikipedia.org/wiki/Electromagnetic_radiation?oldid=682715200 *Contributors:* AxelBoldt, Tobias Hoevekamp, Bryan Derksen, Timo Honkasalo, The Anome, AdamW, Youssefsan, Fredbauder, PierreAbbat, Ray Van De Walker, DrBob, TomCerul, Heron, Olivier, Stevertigo, Lir, Patrick, Tim Starling, LenBudney, Gabbe, Looxix~enwiki, Ahoerstemeier, William M. Connolley, Den fjättrade ankan~enwiki, Julesd, Glenn, Jeandré du Toit, Mxn, Smack, Pizza Puzzle, Reddi, Nv8200pa, SEWilco, Omegatron, Phoebe, EikwaR, Denelson83, Phil Boswell, Donarreiskoffer, Robbot, Hankwang, Agilulfe~enwiki, Blainster, Wikibot, Wereon, Aetheling, Ramir, Enochlau, Srtxg, Wjbeaty, Giftlite, Snags, Peruvianllama, Anville, Bensaccount, Ssd, AJim, Saaga, Bobblewik, Edcolins, Louis Labrèche, Utcursch, Clinton reece, Antandrus, Aulis Eskola, Karol Langner, Rdsmith4, Icairns, Mozzerati, Craig Currier, Jcw69, Buickid, Deglr6328, Adashiel, Canterbury Tail, The stuart, Maestrosync, Mike Rosoft, Discospinster, Guanabot, Pak21, Vsmith, Jpk, CODOR, Mykhal, Kbh3rd, Kaisershatner, Dkroll2, El C, Anphanax, Rgdboer, Hayabusa future, Laurascudder, Edward Z. Yang, Bobo192, Smalljim, I9Q79oL78KiL0QTFHgyc, Nk, Franl, Deryck Chan, Ranveig, Storm Rider, Msh210, Alansohn, Jamyskis, The RedBurn, Atlant, Snowolf, KJK::Hyperion, Wtmitchell, Kdau, RainbowOfLight, DV8 2XL, Akidd dublin, Mpatel, Prashanthns, Graham87, Magister Mathematicae, Abach, Chun-hian, Sjö, Rjwilmsi, JVz, Ian Lancaster, The wub, Bhadani, Oliverkeenan, FlaBot, Michaelbluejay, Ysangkok, Gurch, Fresheneesz, Lmatt, Srleffler, Chobot, DVdm, The Rambling Man, YurikBot, Wavelength, Crotalus horridus, Sceptre, Arado, Bhny, JabberWok, Kerowren, Rintrah, Salsb, SEWilcoBot, Grafen, Jpowell, Killdevil, Scottfisher, Bota47, Kkmurray, Ms2ger, Tigershrike, Light current, Enormousdude, 2over0, C h fleming, Sagsaw, Closedmouth, Zerodamage, Modify, JoanneB, Sizarieldor, GrinBot~enwiki, Serendipodous, Mejor Los Indios, Sbyrnes321, Veinor, SmackBot, Melchoir, Pgk, C.Fred, The Photon, KocjoBot~enwiki, Delldot, Binarypower, HalfShadow, Gilliam, Dauto, Simsea, Rmosler2100, Chris the speller, Keegan, MK8, Complexica, The Rogue Penguin, VincenzoAmpolo~enwiki, Octahedron80, DHN-bot~enwiki, Sbharris, Can't sleep, clown will eat me, JustUser, Chlewbot, Pax85, Shadow1, Drphilharmonic, Hammer1980, Mwtoews, Daniel.Cardenas, Tfl, DJIndica, Nmnogueira, SashatoBot, Lambiam, Vasu123, Tefnut~enwiki, Calum MacÙisdean, JorisvS, Bjankuloski06en~enwiki, Melody Concerto, Jmorkel, Noah Salzman, Alethiophile, Topazg, Rogerbrent, Dicklyon, Macellarius, Doczilla, Jose77, Caiaffa, Smin0, R~enwiki, GDallimore, IanOfNorwich, Jp0186, Rgjm, Tawkerbot2, JRSpriggs, G-W, Chetvorno, JForget, Sakurambo, Armin T, GHe, WeggeBot, Eecon, Bvcrist, FIL (usurped), Raomap, Yeanold Viskersenn, Odie5533, Aajaja, Doug Weller, Christian75, Dchristle, NMChico24, Omicronpersei8, RDates, Thijs!bot, Epbr123, Barticus88, Mbell, Yy-bo, Martin Hogbin, N5iln, Headbomb, KSSA, Marek69, Nick Number, SvenAERTS, Handface, David D., KrakatoaKatie, AntiVandalBot, Luna Santin, Seaphoto, QuiteUnusual, RapidR, Chill doubt, Rico402, Lfstevens, Andreazy, Arx Fortis, Golgofrinchian, IrishFBall32, JAnDbot, MER-C, Andonic, 100110100, Acroterion, Bongwarrior, VoABot II, Jetstreamer, JNW, JamesBWatson, Mclay1, Nyttend, Catgut, Allstarecho, V 1993, InvertRect, RisingStick, Ashishbhatnagar72, NatureA16, Oren0, Hdt83, MartinBot, Mermaid from the Baltic Sea, Rettetast, Federico Benitez Conte, Kostisl, Kateshortforbob, Lcabanel, Nono64, J.delanoy, Troyboy53, Uncle Dick, Javawizard, Kar.ma, Eliz81, Extransit, StalinsLoveChild, Cpiral, St.daniel, Darkspots, McSly, Tarotcards, Gurchzilla, NewEnglandYankee, ARTE, Mufka, Fylwind, Atropos235, Cometstyles, Ibrasg, Sheliak, Sokratesla, Black Kite, VolkovBot, Thedjatclubrock, AlnoktaBOT, Philip Trueman, Yakitoriman, TXiKiBoT, GLPeterson, Paine, Nxavar, Qxz, Seraphim, Bass fishing physicist, Jackfork, Mishlai, WikiCantona, Venny85, Stamulevich, MADe, Lamro, Editorpark, John David Wright, Rhopkins8, Dianneknight, Goodwill289, Bsayusd, Logan, EmxBot, Deconstructhis, Ratsbew, The Random Editor, EJF, SieBot, Spammerman, Demologian, Scarian, WereSpielChequers, Jim E. Black, Jauerback, Gerakibot, Tigerdragon, The way, the truth, and the light, LeadSongDog, Likebox, Flyer22 Reborn, Paolo.dL, JSpung, Antonio Lopez, Hatster301, Dtvjho, Iknowyourider, Thinghy, Mike2vil, Mygerardromance, Dust Filter, BentzyCo, Sbacle, Lascorz, Jlc0023, Rickcandell, ClueBot, LAX, Trojancowboy, Binksternet, Vladkornea, PipepBot, Snigbrook, The Thing That Should Not Be, Amen316, Thubing, Kanhef, CrazieXninja, Catintehbox, Razimantv, Boing! said Zebedee, Ravirathore1984, VandalCruncher, Yongy, Vql, DragonBot, Djr32, Excirial, Jusdafax, Eeekster, Abrech, Rubin joseph 10, SpikeToronto, Brews ohare, PhySusie, M.O.X, Kaiba, Aitias, Jonverve, Amaltheus, CorpITGuy, SoxBot III, Anon126, Editorofthewiki, Jytdog, BodhisattvaBot, Dthomsen8, Nicoguaro, Shieber, SilvonenBot, WikiDao, ZooFari, Shikasannin, HexaChord, Addbot, Eric Drexler, Some jerk on the Internet, Dharmendra srivastva, Tcncv, Betterusername, Ukberry, Ronhjones, TutterMouse, Njaelkies Lea, Fieldday-sunday, Ironholds, Laurinavicius, GyroMagician, Churibo, Redheylin, Epzcaw, AndersBot, Favonian, Doniago, LinkFA-Bot, 5 albert square, Tide rolls, Lightbot, Teles, Meisam, Genius101, Legobot, Luckas-bot, Yobot, THEN WHO WAS PHONE?, KamikazeBot, Linktex, Tempodivalse, AnomieBOT, Jim1138, Galoubet, Bluerasberry, Materialscientist, Legofreak2008, Citation bot, ArthurBot, Xqbot, Phazvmk, Cureden, Romanfall, Capricorn42, Emezei, Nickkid5, The Original Economist, Br77rino, Loiskristellemum, GrouchoBot, Redpanda900, RibotBOT, Victamonn, Www.ca, Mathonius, Maplestory101, Doulos Christos, Sophus Bie, January2009, Shadowjams, Erik9, Kierkkadon, Niaoulibloodelf, Earwax09, Prari, ImaFirinMaLazor, Lookang, Amilnerwhite, Vuldoraq, Steve Quinn, JameKelly, Austria156, HamburgerRadio, Citation bot 1, Alipson, Redrose64, Pinethicket, MJ94, A8UDI, Jschnur, MondalorBot, Merlion444, December21st2012Freak, Jauhienij, Ronak abna, Trappist the monk, Yunshui, Sumone10154, J-p-fm, Reaper Eternal, Suffusion of Yellow, Reach Out to the Truth, Marie Poise, 360flip360, Onel5969, RjwilmsiBot, TjBot, Rollins83, EmausBot, John of Reading, Acather96, Bio watcher, WikitanvirBot, 8lak3st3r, Immunize, Poi830, Racerx11, Yt95, Sp33dyphil, CoincidentalBystander, Tommy2010, Netheril96, Wikipelli, Bdjwww, Thedoctor123, Susfele, 1howardsr1, Danilomath, Quinnd16, OrdinaryFattyAcid, Sky380, Otuguldur, Maxrokatanski, L Kensington, 图图图图~enwiki, Donner60, Zueignung, ChuispastonBot, RockMagnetist, ResearchRave, ClueBot NG, Twihard123, Lcdrovers, Encycloshave, Frietjes, Braincricket, Mesoderm, HazelAB, Marechal Ney, Sina-chemo, Go Phightins!, Widr, MerlIwBot, Diyar se, Helpful Pixie Bot, Calabe1992, Bibcode Bot, DBigXray, Lowercase sigmabot, BG19bot, Ryanross43, Vivek Verma 38, AvocatoBot, Robert the Devil, TROPtastic, Sparkie82, YVSREDDY, Zedshort, Shaun, Tomohama, BattyBot, LeviathanPMS, ChrisGualtieri, BlazeNinja418, Khazar2, MSUGRA, Dániel I fiz, Gdrg22, JYBot, TopGarbageCollector, Agumonkey, BrightStarSky, Dexbot, Tharvey100, AyaLovesAmjad, Webclient101, CuriousMind01, Epicgenius, Sɛvɪnti faɪv, Mrsquirrel dh, Teeth69, Kharkiv07, Ugog Nizdast, Jordanvandijk, DavRosen, Hanthoec, Jianhui67, SpecialPiggy, Param Mudgal, Linuxjava, CharlesIJ1948, Lakun.patra, Csutric, Monkbot, Startellar, Gauracho, Vieque, Joeleoj123, Quarter2002, Nelsonaugust3, Tris1313, Indranil1993, WyattAlex, Anmikmore, Easy Secrets, SageGreenRider, Electric Toast, Telkomsel013, Username123123123, KasparBot, Chrisemblhh, Loudandrews123, Lantolar and Anonymous: 884

- **Electromagnetic radiation and health** *Source:* https://en.wikipedia.org/wiki/Electromagnetic_radiation_and_health?oldid=686820675 *Contributors:* Timo Honkasalo, Europrobe, Heron, Rsabbatini, LenBudney, Aarchiba, Floydian, Jeffq, Twang, Hankwang, Giftlite, Solipsist, Golbez, Beland, Jmeppley, Qui1che, Arcataroger, Discospinster, Rich Farmbrough, BBUCommander, Vsmith, LindsayH, Bender235, Neko-chan, Enric Naval, Congruence, DreamGuy, Wtshymanski, Kdau, Gene Nygaard, Martian, Bobrayner, Awostrack, Orangehues, AdinaBob, ObsidianOrder, Edison, Rjwilmsi, FlaBot, Michaelbluejay, Ground Zero, Srleffler, DVdm, Wavelength, Chris Capoccia, Hydrargyrum, Perry Middlemiss, 2over0, Chase me ladies, I'm the Cavalry, Willy zilly nilly, Mossig, Crimsondestroyer, Liujiang, SmackBot, CyclePat, Ohnoitsjamie, Chris the speller, Bazonka, Colonies Chris, Милан Jелисавчиh, Frap, Cícero, Cybercobra, Bradenripple, Topazg, Geologyguy, Arathalion, Alessandro57, Hyperman 42, Fernvale, MightyWarrior, Alexander Iwaschkin, DJPhazer, VoxLuna, DangerousPanda, IntrigueBlue, Prosthetic Head, Dongbong, Bob Stein - VisiBone, Thijs!bot, Chris01720, Second Quantization, Guy Macon, Tjmayerinsf, JAnDbot, Txomin, The Transhumanist, Acroterion, JamesBWatson, Destynova, BrianGV, WhatamIdoing, Cgingold, A3nm, Ppeltola, Markus451, Anaxial, Sm8900, Paunaro, Astro-Hurricane001, Courage Dog, DadaNeem, Fagiolonero, Someguy1221, Dianneknight, Doc James, Buckyrogers, Biscuittin, StAnselm, LeadSongDog, Flyer22 Reborn, Alexbrn, Wdwd, ClueBot, PipepBot, ImperfectlyInformed, Rivkah77, Scohen2401, Detroiterbot, Ckenyon3449, Jonverve, Kozzz, XLinkBot, DOI bot, Simonm223, GeoffreyBanks, Atethnekos, Ronhjones, TutterMouse, Juanpablosoto, Verbal, Lightbot, Pensees, OlEnglish, Jarble, QueenCake, AnomieBOT, Nutriveg, DGaryGrady, Materialscientist, Wodawik, Citation bot, Pontificalibus, RG72, Rbassilian, FrescoBot, Voxii, Jc3s5h, CapitalElll, Citation bot 1, Crusherdust, Gmenta, Orenburg1, Trappist the monk, Lotje, Quintupeu, H2eaux, DARTH SIDIOUS 2, Dfenelus1, User8253, RjwilmsiBot, Jojo129~enwiki, GoingBatty, Yourworld07, TacitSilence, Alpha Quadrant, ClueBot NG, Faizanalivarya, Horoporo, Widr, Geofferybard, Helpful Pixie Bot, Bibcode Bot, Neøn, Safeemf, Altaïr, Vladislav Oleshchenko, JMtB03, 220 of Borg, BattyBot, Geneskl, APerson, Dexbot, JZNIOSH, TippyGoomba, Fmc47, Faizan, Comp.arch, A Certain Lack of Grandeur, Mfb, Dongseong Hwang, Drkvncnt, Jasonlee8985, Monkbot, JohnWhoMeansWell, Bandsvase1, Thadlooms32, Jointdolls565, GH342, Ninjabeard99, Danceever1, Emrhealth, Donaberyl6, Mario Castelán Castro, Dickbrettbohrer, Knaveknight and Anonymous: 152

- **International Agency for Research on Cancer** *Source:* https://en.wikipedia.org/wiki/International_Agency_for_Research_on_Cancer?oldid=688061149 *Contributors:* Mac, Jfdwolff, Gbr3~enwiki, Evand, Axl, Alai, Drbogdan, Rjwilmsi, Linuxbeak, Ground Zero, Physchim62, RussBot, Albedo, Crystallina, FocalPoint, Stepa, Sveika, Tisthammerw, Cybercobra, Vina-iwbot~enwiki, Cattleyard~enwiki, WeggeBot, Hemlock Martinis, Ebyabe, Faigl.ladislav, Escarbot, WinBot, Dogru144, Magioladitis, Jimjamjak, Cgingold, Pvosta, SieBot, Rystheguy, ImperfectlyInformed, PixelBot, Muro Bot, Jonverve, HannahRedmond, Good Olfactory, Addbot, Devadatta, Luckas-bot, Yobot, Shootbamboo, ArthurBot, Lulu97417, Andromeas, Ttweed, Jesse V., WikitanvirBot, ZéroBot, COM-IARC, MusikAnimal, MrBill3, Polmandc, Creosota, Comp.arch, Jameskirk, Monkbot, 115ash, Pbinlyon and Anonymous: 16

- **List of IARC Group 2B carcinogens** *Source:* https://en.wikipedia.org/wiki/List_of_IARC_Group_2B_carcinogens?oldid=675045464 *Contributors:* Cherkash, Giftlite, Eequor, Chowbok, Anodyne, Reinyday, Cmdrjameson, Arakin, Pixeltoo, BD2412, Physchim62, CambridgeBayWeather, Joel7687, Open2universe, FocalPoint, Shoy, Edgar181, Gilliam, Bluebot, Guroadrunner, Javsav, Cgingold, I.Scream~enwiki, Cobi, LeadSongDog, ClueBot, Maxcip, AlptaBot, Excirial, Jonverve, SDY, Addbot, Tassedethe, AnomieBOT, Nutriveg, Paul Bedson, Jsjsjs1111, Operative67, 570ces, Morineau, Comp.arch, Nerppy, VirtuOZ and Anonymous: 15

- **Precautionary principle** *Source:* https://en.wikipedia.org/wiki/Precautionary_principle?oldid=685691623 *Contributors:* Lee Daniel Crocker, Eloquence, Bryan Derksen, Robert Merkel, The Anome, Ed Poor, Rmhermen, Anthere, Heron, B4hand, Bobdobbs1723, Boud, Michael Hardy, Alfio, Snoyes, Bogdangiusca, Nikai, Corixidae, Kaihsu, Palfrey, Astudent, º¡º, Wik, Bhuston, Populus, Modulatum, Wolfkeeper, Crasch~enwiki, No Guru, Arnejohs, Guanaco, Kravietz, JRR Trollkien, Sam Hocevar, Pgreenfinch, Neutrality, Punchi, Flex, Freakofnurture, Spiffy sperry, Rich Farmbrough, Vsmith, ArnoldReinhold, Westendgirl, Ttguy, PhilHibbs, Dtremenak, Vortexrealm, AppleJuggler, Pearle, Paul Bonneau, Arthena, Rd232, Ricky81682, Apoc2400, Mailer diablo, Ynhockey, Batmanand, Sponge, Snowolf, Brownpau, Alai, Drbreznjev, Bookandcoffee, Kazvorpal, Woohookitty, Carcharoth, Behun, Graham87, Parmaestro, David Levy, Rjwilmsi, Pdelong, XP1, NeonMerlin, Mahlum~enwiki, Krueschan, Jrtayloriv, Srleffler, Common Man, WhyBeNormal, GangofOne, Sasoriza, Dj Capricorn, FrankTobia, YurikBot, Wavelength, RussBot, PWhittle, Tralala~enwiki, Tresckow, Chaser, Dysmorodrepanis~enwiki, Mosquitopsu, Daniel Mietchen, Epipelagic, Mike Treder, Wknight94, Closedmouth, Arthur Rubin, Rob G Weemhoff, Allens, Emcee, Sardanaphalus, SmackBot, Tobias Schmidbauer, Reedy, Unyoyega, Mgreenbe, Miguelaznar, Ohnoitsjamie, Hmains, Angelbo, Chris the speller, RDBrown, Colonies Chris, Jon Nevill, Argyriou, "alyosha", Someoneisatthedoor, Richard001, Metamagician3000, DDima, StN, Arnoutf, Mringgaard, Bcasterline, Arodb, Staalmannen, Anlace, Wickethewok, AT2663, Ckatz, Bbold, Hu12, Apathos, Beherbert, Dartelaar, Eastlaw, Dmatisoff, CmdrObot, Argon233, Knappster, Woffie, Tawkerbot4, Chris Henniker, Japanman, Getf*cked, Electron9, Ajkr925, Ben Harris-Roxas, Lfstevens, Samuel Erau, Aavrakot, Epeefleche, Albany NY, Prospect77, PhilKnight, Rosie.cooney, Ralgara, Father Goose, WhatamIdoing, Jimjamjak, Hbent, DGG, Jim.henderson, Nijhofrene, Wild Pansy, R'n'B, Reblf, Maurice

Carbonaro, Skumarlabot, DanaJayne, Skier Dude, Tarotcards, Jorfer, Dunsandel, VolkovBot, Johnfos, Gonzeaux, Rmauger, Thadius856AWB, Raymondwinn, Pnprice, Pvednes, Infineede, SylviaStanley, Hrafn, Phe-bot, WRK, Nopetro, PhilMacD, Danelosis, Ricklaman, Der Golem, Eiland, DerekMorr, ComputerGeezer, Uttaranchal, Frozen4322, SchreiberBike, Polly, The Baroness of Morden, Jytdog, Dark Mage, SilvonenBot, Winged Cat, Addbot, Pensatrice, Jncraton, Pince Nez, Yobot, A.k.a., AnomieBOT, Bsimmons666, Materialscientist, Juanita09, Citation bot, Xqbot, Account3915, DSisyphBot, GrouchoBot, Dzsi, MerlLinkBot, Vladimir.frolov, FrescoBot, Citation bot 1, Micromesistius, Faucnet, Digenti, Sgsg, Informed counsel, RjwilmsiBot, Pmiddlemas, Solomonfromfinland, Tolly4bolly, Scientific29, Rangoon11, ChuispastonBot, Teapeat, Teaktl17, ClueBot NG, Vergilden, Chester Markel, CitationCleanerBot, Awsmkid101, MrBill3, RealZero~enwiki, IjonTichyIjonTichy, Mogism, ClutchTheMagnum, Lhays, Lalo3767, BrilliFAN, Intlpol123, Caligo57, 4Truth2020, Eddu1973, Monkbot, Jennchiu, راسيسي, Aniroodh Sarkar, Jerodlycett and Anonymous: 176

- **Scientific Committee on Emerging and Newly Identified Health Risks** *Source:* https://en.wikipedia.org/wiki/Scientific_Committee_on_Emerging_and_Newly_Identified_Health_Risks?oldid=616263256 *Contributors:* R'n'B, Int21h, Deselliers and GabeIglesia

- **Wireless electronic devices and health** *Source:* https://en.wikipedia.org/wiki/Wireless_electronic_devices_and_health?oldid=686603055 *Contributors:* The Anome, Heron, Rsabbatini, Ronz, Timwi, Alan Liefting, Solipsist, Qui1che, Hydrox, QuantumEleven, Elpincha, Nick Moss, Wtshymanski, Bobrayner, Mindmatrix, ObsidianOrder, Rjwilmsi, Vegaswikian, Nick mallory, Bmicomp, Shaddack, Badagnani, Retired username, Mikeblas, 2over0, Chase me ladies, I'm the Cavalry, Mossig, Twilight Realm, SmackBot, Gilliam, Kmarinas86, Bluebot, Sawran~enwiki, Aubray1741, Fiskbullar, Evolve2k, AThing, William Wang, JHunterJ, Topazg, Arathalion, Hyperman 42, Fernvale, Oden, Cydebot, Grand Dizzy, Balint256, Marokwitz, Tjmayerinsf, Niaz, JamesBWatson, Cgingold, Sm8900, Tskam1, Kateshortforbob, Silverxxx, LordAnubisBOT, Funandtrvl, AlleborgoBot, ClueBot, ImperfectlyInformed, Socrates2008, Unprovoked, Dank, Jonverve, Canberra photographer, RDOlivaw, XLinkBot, OhioTrivium, AndreNatas, Addbot, GeoffreyBanks, Axtelldm, Fgnievinski, Verbal, Pensees, Yobot, AnomieBOT, Citation bot, RibotBOT, Jeremystalked, RjwilmsiBot, Jalusbrian, John of Reading, DMAch, Slightsmile, Aarp65, Mariuskempe, Blessingsncheers, Helpful Pixie Bot, Onoclea, DBigXray, Frze, Lachie h, Happenstancial, TippyGoomba, Catch2424, BurritoBazooka, Marcusknight23, T.Giffords, Monkbot, Seosavants, CaseyMillerWiki and Anonymous: 63

3.2 Images

- **File:2007Computex_e21Forum-MartinCooper.jpg** *Source:* https://upload.wikimedia.org/wikipedia/commons/1/1f/2007Computex_e21Forum-MartinCooper.jpg *License:* CC-BY-SA-3.0 *Contributors:* Rico Shen *Original artist:* Rico Shen

- **File:Active_mobile_broadband_subscriptions_2007-2014.svg** *Source:* https://upload.wikimedia.org/wikipedia/commons/0/05/Active_mobile_broadband_subscriptions_2007-2014.svg *License:* CC BY-SA 4.0 *Contributors:* Own work *Original artist:* Chris55

- **File:Ambox_current_red.svg** *Source:* https://upload.wikimedia.org/wikipedia/commons/9/98/Ambox_current_red.svg *License:* CC0 *Contributors:* self-made, inspired by Gnome globe current event.svg, using Information icon3.svg and Earth clip art.svg *Original artist:* Vipersnake151, penubag, Tkgd2007 (clock)

- **File:Ambox_important.svg** *Source:* https://upload.wikimedia.org/wikipedia/commons/b/b4/Ambox_important.svg *License:* Public domain *Contributors:* Own work, based off of Image:Ambox scales.svg *Original artist:* Dsmurat (talk · contribs)

- **File:Atmospheric_electromagnetic_opacity.svg** *Source:* https://upload.wikimedia.org/wikipedia/commons/3/34/Atmospheric_electromagnetic_opacity.svg *License:* Public domain *Contributors:* Vectorized by User:Mysid in Inkscape, original NASA image from File:Atmospheric electromagnetic transmittance or opacity.jpg. *Original artist:* NASA (original); SVG by Mysid.

- **File:Cellphone_aerial_mast.jpg** *Source:* https://upload.wikimedia.org/wikipedia/commons/b/b5/Cellphone_aerial_mast.jpg *License:* CC-BY-SA-3.0 *Contributors:* photo taken by Harald Hubich *Original artist:* Harald Hubich

- **File:Commons-logo.svg** *Source:* https://upload.wikimedia.org/wikipedia/en/4/4a/Commons-logo.svg *License:* ? *Contributors:* ? *Original artist:* ?

- **File:DynaTAC8000X.jpg** *Source:* https://upload.wikimedia.org/wikipedia/commons/7/74/DynaTAC8000X.jpg *License:* CC BY-SA 3.0 *Contributors:* http://en.wikipedia.org/wiki/File:DynaTAC8000X.jpg *Original artist:* Redrum0486

- **File:EM_spectrum.svg** *Source:* https://upload.wikimedia.org/wikipedia/commons/f/f1/EM_spectrum.svg *License:* CC-BY-SA-3.0 *Contributors:* ? *Original artist:* ?

- **File:Electromagneticwave3D.gif** *Source:* https://upload.wikimedia.org/wikipedia/commons/4/4c/Electromagneticwave3D.gif *License:* CC BY-SA 3.0 *Contributors:* Own work *Original artist:* Lookang many thanks to Fu-Kwun Hwang and author of Easy Java Simulation = Francisco Esquembre

- **File:Electromagneticwave3Dfromside.gif** *Source:* https://upload.wikimedia.org/wikipedia/commons/a/ad/Electromagneticwave3Dfromside.gif *License:* CC BY-SA 3.0 *Contributors:* Own work *Original artist:* Lookang many thanks to Fu-Kwun Hwang and author of Easy Java Simulation = Francisco Esquembre

- **File:Emblem_of_the_United_Nations.svg** *Source:* https://upload.wikimedia.org/wikipedia/commons/5/52/Emblem_of_the_United_Nations.svg *License:* Public domain *Contributors:* Based on File:Flag_of_the_United_Nations.svg *Original artist:* Spiff

- **File:FarNearFields-USP-4998112-1.svg** *Source:* https://upload.wikimedia.org/wikipedia/commons/5/5d/FarNearFields-USP-4998112-1.svg *License:* Public domain *Contributors:* US Patent 6657596 *Original artist:* Goran M Djuknic

- **File:Flag_of_Europe.svg** *Source:* https://upload.wikimedia.org/wikipedia/commons/b/b7/Flag_of_Europe.svg *License:* Public domain *Contributors:*

- File based on the specification given at [1]. *Original artist:* User:Verdy p, User:-xfi-, User:Paddu, User:Nightstallion, User:Funakoshi, User:Jeltz, User:Dbenbenn, User:Zscout370

3.3   Content license